I Have Lived Four Lives

I HAVE LIVED FOUR LIVES...

Wilfred Buck

ARP Books | Winnipeg

ARP Books (Arbeiter Ring Publishing)
205-70 Arthur Street
Winnipeg, Manitoba
Treaty 1 Territory and Historic Métis Nation Homeland Canada R3B 1G7
arpbooks.org

Cover design by Bret Parenteau.
Interior layout by Relish New Brand Experience.
Cover image detail from *Ininew Achakosuk Masinikan* by Annette S. Lee,
William Wilson, & Wilfred Buck.
Printed and bound in Canada by Kromar Printing on certified FSC® paper.

ARP Books acknowledges the generous support of the Manitoba Arts Council and
the Canada Council for the Arts for our publishing program. We acknowledge
the financial support of the Government of Canada and the Province of Manitoba
through the Book Publishing Tax Credit and the Book Publisher Marketing
Assistance Program of Manitoba Culture, Heritage, and Tourism.

Library and Archives Canada Cataloguing in Publication
Title: I have lived four lives... / Wilfred Buck.
Names: Buck, Wilfred, author.
Identifiers: Canadiana (print) 20210227737 | Canadiana (ebook) 20210227877 |
    ISBN 9781927886496 (softcover) | ISBN 9781927886502 (ebook)
Subjects: LCSH: Buck, Wilfred. | LCSH: Astronomers—Manitoba—Biography. |
    LCSH: Indigenous peoples—Manitoba—Biography. | LCSH: Dreams. |
    LCSH: Visions. | CSH: Indigenous teachers—Manitoba—Biography. |
    LCGFT: Autobiographies.
Classification: LCC E99.C88 B83 2021 | DDC 971.27004/973230092—dc23

In memory of those that have gone before...
We will try and bring your stories forward...

Prologue   9

Preface   12

Introduction   20

I   38

II   50

III   79

IV   100

Acknowledgements   140

Prologue

Writing in the middle of a pandemic.

The world has gone crazy!

It all started like this…

Connie (my wife), Colin (my son) and myself were at an education conference in Edmonton, Alberta, when it was announced that all congregations of 250 people or more were hereby cancelled! It was March 12th, 2020 and we were told to go home immediately!

We thought nothing of this. If we did it was, "what an over-reaction!" Who in the hell is in charge and making these crazy-ass decisions? What they going to do next? Shut everything down??

Little did we realize what we were in for!

As we were travelling back from Alberta after the big announcement, my oldest daughter Carrie texted us and asked us to stop at a store along the way—and buy toilet paper and hand sanitizer??

She said everywhere her and Mistawasis (our youngest son) went in Winnipeg there was no toilet paper or hand sanitizer. Toilet paper! We did as we were asked and stopped at ten stores. Ten stores! Looking for toilet paper, with no luck! People were going batshit crazy… running around the stores loaded with toilet paper and hand sanitizer… so we joined in on the madness… but we were too late! No luck for us!

When I arrived home and went to work the following Monday, March 16th, again I was told to go home. Self-isolate I was told. Work from home I was told. My first question is how do I self-isolate when there are seven of us staying in my house? My mother, who is eighty-six years old and all four of my children, as well as my wife and I.

That was nineteen weeks ago! We are still alive… well… so far, and stocked with all kinds of canned goods, TOILET PAPER, hand sanitizer, paper towels. My wife has made all kinds of face masks and we managed to get a hold of disposable gloves.

We are lucky! Both my wife and I are educators and thus have NOT been laid off. Our oldest daughter is a home care provider to the elderly and we are always worried about her but she now only sees one client at one residence.

A lot of people have been laid off and are having a very hard time.

We have been trying to help as best we can by assisting families and the homeless… volunteering to cook soup and bannock on a weekly basis.

I have been doing other things like exercising. I've been walking at least eight kilometers a day!

I also began making ceremonial objects again and just this morning realized I could be writing!

Nineteen weeks have gone by and it is only NOW I realize I could have been writing!! Whiskey Tango Foxtrot?

My wildest dreams never prepared me for this! I just got glimpses of lineups and empty shelves… with no context…

We seem to be existing in a surreal reality where shadows of memories run down empty dark streets slipping in and out of the light and darkness.

Memories… *We did this at one point… I remember doing this with all of the family, with all our relatives…*

I wonder, if in three generations, how we interact now will have become "tradition" for our relatives in the future.

As an Elder once said to me, "We fall down and have a decision to make. Do we lay there and feel sorry for ourselves or do we rise up, dust ourselves off and continue our journey towards the future? Either way the future will come, but will it arrive at your doorstep an unwanted guest or as a lesson learned?" Choices...

I will turn sixty-six within the next few days. Never thought I would be allowed to get this far! It has been an awesome ride.

At times dreams have visited me with frequency and clarity and at other times they just gave me glimpses, out of context, with no connection to anything... until they happened in my waking life. Then I would say, "Oh, that is what that dream was about!"

Pawamiuk (dreams) run central in the lives of my people... ININEWUK (the Crees) and I think for any First Nation person here in this sacred land.

These are some of the dreams that were given me. I am the Dream Keeper...

Preface

I am Pawami niki titi cikiw—the Dream Keeper. I am Ininew from Opaskwayak Ininewuk. Ininew is how my people refer to themselves. Ininew (singular) or Ininewuk (plural) refers to the phrase MIXING OF FOUR. It derives from the root word newo which is the number four. Thus we are a mixture of four: earth, air, water, fire; or: physical, emotional, mental, and spiritual; or, four parts: Plains Cree, Woodland Cree, Swampy Cree, and Rocky Cree.

In "The Pilgrim," the great poet Kris Kristofferson wrote of a man whose life is a mix of truth and fiction. He's spent his life trying to run from his darker impulses but he always finds "he's traded in tomorrow for today," and he's lost what he loved. His path has been a contrary mix of evil and good. It seems he's referring to me and those of my ilk. This is my reality—fiction. What you are about to picture here is the artist painting a clean, biased, slightly sullied version of a truer reality. I put the best me on display. I crawl out of the darkness, into the light…

I am also known as Wilfred Alfred Buck, a name given to me by my parents and grandparents.

In my younger days, to the people of my area, I was commonly known as Smokey Buck.

I am of the fresh-out-of-the-bush, partly civilized, colonized, displaced, confused, angry people. Trained and shamed by teachers, preachers, doctors, nurses, law enforcement officials, movies, radio, and television programs to be a pill-popping, hard drinking, self-loathing, easily impressed, angry, non-conformist, mal-adjusted, disaffected youth of the "dirty-indian" baby boomer generation.

I lived, died, and lived again, in North Central Manitoba.

This story talks about the four stages of experience I underwent to get from the past to the present, from here to there, then from there to here all the while learning about "dis, dat and da udder ting" (this, that and the other thing). Believe me, I know a little something about a little something. So here are some of the teachings forced upon my stubborn mind, that hopefully will assist others on their journey of discovery and healing.

Looking back at all the situations I found myself in, death has been watching me and reminding me that it is always around ready to pop up at any time. I am sure if anyone looks back at the places and times when they came close to the end, they are reminded very intimately of their own mortality and the gift of life.

For myself I feel and see a car running over me as I try to "bumper shine" on a cold winter morning… a semi-trailer whizzing by my head as I lay passed out, in a drunken stupor, in the middle of a highway… breaking through the ice on a winter road and fighting to crawl back up onto solid footing as I swallowed water on a cold February night… walking through a blizzard with my clothes frozen solid around my body… going to sleep in a raging blizzard, tired from exhaustion, cold, and frustration… overdosing on LSD and getting lost in the northern boreal forest…

By all rights, I am told, I should be dead many times over. Here is the first of many examples of this…

When I was eight years of age, living in the town of The Pas, Manitoba, I went swimming with a few friends in the Saskatchewan

River. One of our party did not know how to swim and did not think it important to tell us this fact. We will call him Rodger. Rodger was a cousin with someone in our circle. He was visiting from northern Manitoba. From what I understood at the time, Rodger's family was chased out (displaced) from their homes because flooding was to take place on their fathers' hunting, fishing, and trapping territory, thereby disrupting the peoples' livelihood, way of life, culture, and existence.

They were flooded out. They had no way to make a living anymore… thus Rodger's family moved to The Pas in search of other means of existence.

Rodger's family was just one of thousands of families being displaced all along traditional waterways and lakes in northern Manitoba and Saskatchewan during the span of time from the 1950s to the 1990s.

It was a fine sunny July morning and all was right with the world. We arrived at the government dock where we always swam. The dock was built by the Department of Fisheries with the main purpose of supplying communities up and down the Saskatchewan River from Cumberland House, Saskatchewan upstream to Moose Lake and Easterville, Manitoba down towards the big dam at Grand Rapids. Large boats and barges would dock here and load up with all sorts of equipment and supplies destined for the dams and communities up and down the river.

Our newly-acquired buddy (Rodger) decided to dive right into the river… no hesitation! The river where the dock was and where we swam was about 8–9 feet deep and in Rodger went. I dove in right after him and when I broke surface beside him, he was panicking, screaming he could not swim and thrashing about with terror in his eyes!

In his hysterical panic Rodger reached out and grabbed the first thing he saw, which of course was me! He decided to climb on my shoulders in his panic, screaming and yelling as well as pulling my hair, trying to grab on to something so that he could stabilize himself. All this at my expense! Down I went! I was down so deep I could feel the mud from the bottom of the river on my feet. I fought my way back to the surface only to be smothered and submerged again by a maniac who was trying not to drown. Down I went again… coughing and swallowing water. I began to panic… clawing at the person who was standing on my shoulders holding me underwater. As this was happening I began to think, "hey I can touch bottom. If I walk a few steps forward I will be at the dock." I calmed down and began to move forward. I grabbed on to the dock and as I did this the person on my shoulders realized he too could grab the dock. I was saved! I emerged from the water, sputtering, coughing and shivering… but alive. We all laughed… because we would not cry…

Swallowed a lot of prairie silt that day!

This was the first but not the last time I came face to face with death.

I have had the good fortune to travel and speak with people from all over the world. The Indigenous peoples I have shared with tell similar stories and have parallel if not similar experiences. These stories concern colonial oppression, marginalization, poverty, unemployment, violence, addictions, relocation, rape, murder, and genocide.

They also include the loss of natural resources through massive hydroelectric development projects, natural resource extraction such as timber, pulp and paper, freshwater fishing, mining, and development of railways, highways, and power lines.

I was given the opportunity to attend the 1992 Rio Earth Summit held in Rio de Janeiro, Brazil. There were Indigenous people from all over the world there and we were lucky enough to connect with many of them. I say "we" because I was in Rio with a group of us from Canada generally and Manitoba, specifically.

Larry Morrissette, Judy Da Silva (nee Williamson) and I were part of the Manitoba contingent which was part of a larger Canadian Indigenous contingent. Rio! Larry and I were very excited to go to the Copacabana Beach because we'd heard so much about it.

A couple of things I noticed while in Brazil was that the sun made a curious path across the sky. I also noticed that the night sky was unfamiliar to me!

Back home in Canada, *Pisim*, the sun, rose in the east went across the sky in a Southerly arc and set in the west. There in Brazil the sun rose in the east went across the sky in a Northerly direction before setting in the west. *Tipiskawi Pisim* (moon) did likewise.

I was given a shock when I had the opportunity to view the night sky… no big dipper, north star, little dipper! Alien skies! I was under an alien sky!

I talked with some locals and they said there was no pole star like we had in the northern hemisphere but they used a group of stars called "Cruz do sul," the Southern Cross, to point them to the south pole or south direction.

While in Rio and surrounding event sites, we had the good fortune of meeting and speaking with people from all over the world… Indigenous people. We met Sami from northern Finland, Sweden, Finland, and Russia. We met Zapotec from southern Mexico, Kiyapo from southern Brazil, Yanomami from the Amazon Basin, Zuni from New Mexico, Aborigine from Australia, Maori from New Zealand, Ainu from northern Japan, Quechua from Peru… to name a few of the awesome people we connected with!

We went to a few official gatherings while in Rio and I chanced to meet singer John Denver and actor Edward James Almos. They were very nice, soft-spoken people.

During this time, a Japanese journalist interviewed me. She said she was working with street youth in her country and liked what she heard me say. She asked about our Canadian youth and particularly our First Nations urban youth. She asked questions for about two hours and when she finished listening to what I had to say, I was surprised to see she had tears in her eyes. She told me I should write my experiences down... "it could help the youth," is what she said. At the time I thought about it then let it pass. She was not the first person to suggest I write a book.

Upon hearing of some of my personal adventures, many people tell me I should share these with the world, especially with our youth. I was told that I should write a book about my experiences on this earth. I tell these people that I will wait for a "sign."

For me, the "signs" I wait for come in the form of dreams. Pawamiuk. Dreams, that give me glimpses of an event in progress and when I wake up I tell my wife I had such and such a dream—I wonder what that's all about? A good number of times these dreams play themselves out in reality at some point down the path.

I know all this ties in with the name the ancestors gave me, Pawami niki titi cikiw... the Dream Keeper. Dreams are very important to my people. They are part of this process and influence the decisions I make. They are central to who I am and what I do. I am Pawami niki titi cikiw—the Dream Keeper—and dreams guide, heal, direct, show, comfort, warn, and offer possibilities.

I have dreamed about planes crashing into buildings, asteroids punching through the cloud cover and exploding. I have dreamed of explosions and big waves across a great expanse of water.

I have dreamed of ceremony and of giving spirit-names to rows of people. I have dreamed of colours and unrecognizable locations. I have dreamed of people and they have come to me. These dreams are the "signs."

The dreams reveal themselves in their own way... in their own time.

Here is the way one such dream sequence began.

The one dream that told me it was time to begin to write a book was this:

In this dream I am travelling and arrive somewhere… I do not know. I am walking along a small lake with another person. We (another person who I do not know) are sitting by a small body of water looking out across to the other shore, talking. The sun is shining and everything has a golden hue to it (I know this is significant because I have had dreams like this in the past and the golden hue usually signifies milestone markers along the travelled path). When we are done talking, this person (it turns out to be a journalist from London, England) says "you should write a book." The dream ends here. I wake up and tell my wife about this and say, "Wonder what that's about?"

A few months down the road finds me in Jasper, Alberta, presenting at the annual Jasper Dark Sky Festival. As part of the festival, journalists from all over the world are there covering the event and are interviewing presenters. One gentleman approaches me and passes tobacco to me and asks if he could speak with me. I am interested because he has taken time to ask someone about protocol when approaching a traditional First Nations person and has brought tobacco. I agree to sit and talk with him.

We walk along a path which ends up by a small lake and take a seat that is near the path by the lake. The sun is beginning to set over the mountains and everything is covered in a golden hue. "Humm" I say to myself, "where have I seen this before?" A déjà vu moment. We finish the interview and the gentleman gets up and begins to walk away. He stops, turns around and says to me, "You should write a book. You have lived four lives and you should write a book about it." With that he walks away and I never see him again.

I am left standing there… stunned! My dream comes rushing back to me in all its reality and I also remember a time back in 1991 when I began writing some of my experiences down in short story form. To save them on the computer I had given them the heading *I Have Lived Four Lives*!

My dream materializes and I come to the understanding that I guess I should begin the process of writing a book.

A follow-up to this happened a short while later in a dream concerning my children. I was blessed with the care of four amazing spirits: She-Bah-Gi-Shi-Goo Inini, Under The Sky Man also known as Colin; Pinesew Neepahwew Muskysee, Thunderbird Standing Strong, aka Carrie-Marie; Mistawasis, Big Child (a noted Cree Chief at the time of Poundmaker and Big Bear); and Tukwakin Nootin, Autumn Wind, our baby. Two boys and two girls. These ones, along with my wife Muskwa Atchak Iskwew, Bear Spirit Woman aka Connie, are my reason for waking up in the morning and staying sane each day, although some would argue with the sane part.

The dream saw my children gathered round in a large room or meeting place talking. They were sharing stories of some of the things we did together… sitting around… laughing. I can hear them talking and laughing and I got the feeling I had to leave something behind for them. When I woke up, the first thing that was in my mind was a book… I had to write a book.

Being the stubborn bear that I am I had been resisting the message, but finally I got the gist and surrendered. OK, OK I will begin to write…

The question I had was, not if I *should* write a book but rather, *could* I write a book? I have read extensively and what I have read has made me think, feel, and question. Could I write something that could make people think, feel, or question?

What do you think?

Hey… maybe I could write a book!

This narration will tell you what has happened at various times in my life… the ups and downs… highs and lows… the ebbs and flows. So… here it goes…

Introduction

It seems that the years will stretch on forever... from the perspective of youth. With the gaining of patience and age those years tend to shrink drastically. First there is too much time... then not enough...

Memories in time. Memories embedded in emotion, in spirit, in *Mitay*—the heart. These are the learning paths of my people.

I have dreamed this. I have seen this. I have experienced this. I have felt this.

...now I will remember...

I remember, in my very early youth, sitting by the river on a warm September night, staring up at the unending sky... wondering... wondering... wondering. The river was big, the forest bigger, but the sky, the sky was incomprehensible. The age old questions... who are we and what is our place in this reality?

I have come to learn my people have a very different understanding of what "reality" is. There is a term, *miswa*, which could translate into the phrase "all that is." Taken at face value this phrase means everything... but WHAT does everything MEAN?

As Ininew we have NO idea what "everything" means. Everything is incomprehensible... unfathomable. We just know it IS.

We are given to understand our Creator thought miswa into being... all that is and Creator is incomprehensible thus Creation is incomprehensible. When Creator is thought about or discussed, there is no immediate association with a "god head" sitting in the heavens...ruling over all. For the Ininewuk, Creator is an indescribable thought. *Tapwiy Mamascatch* our people would say... totally awesome! But the phrase is also used as a sarcasm when belittling someone.

We are given to understand Creator is omnipotent... never ending. If this is the case then that essential part of us that has been made by our Creator, *Achak*... our spirit... is never ending. This physical place we find ourselves, *oota Aski* (this earth) is just a blink in the greater span of things. We are here for a short visit. Spirits learning how to be human.

The reasoning goes on to say that if Creator is omnipotent then Creator's thought processes are now, as we read this, still in process. Creation is still being created. It is expanding because Creator never stops thinking! So when we look at a firefly in the night, a flash of lightning, a dust particle floating in the breeze, the vastness of the night sky... we realize how immense "all that is" truly is. We come to terms with the concept of how little we know. *Nitimakiin...* I AM PITIFUL.

Our memories connect some of the dots in this great mystery...

My first recollections are of the river, *Kisiskâciwan Sipi...* the Saskatchewan River, as it made its way across Manitoba from West to East to end at the Grand Rapids and then emptied into Lake Winnipeg.

I remember the Saskatchewan River and all the times we spent playing on its muddy banks, swimming in its muddy waters and fishing in its silt heavy soup. Funny thing was in the winter we would drink from the river since the erosion process was reduced to a minimum and the water was clear.

In the language of my people, the *Ininew*, one of the words for the color blue is sipikook, referring to the reflection of the sky on the water. This tells someone listening that the river ran clear... to have the ability to reflect the sky. This word tells us that prior to the extraction of *amisk*, the beaver, from the local environments... ALL OVER North America... a lot of rivers ran clear.

The Fur Trade fueled the push for beaver pelts. Fashion demanded FELT for the style of the time... HATS. Felt hats were a rage in Europe for almost three centuries... three hundred years.

The quest for fur changed the economy and worldview of the Indigenous people of the Americas. The colonization process was introduced...

I read in one of the Hudson Bay journals pertaining to the Fur Trading post at Cumberland House on the Saskatchewan River that one million beaver pelts were extracted from the area in one season. Now you look at this and expand this number to all fur trading companies in all regions of North America and include over two centuries of harvesting; then you have at various seasons like spring, vast areas under spring run-off flood conditions, unchecked by beaver dams, due to the beaver being removed from the environment. The dams the beaver built and maintained held back spring run-off waters, kept erosion at a minimum, controlled water temperatures and provided green shoots for plant-eating animals. With the removal of the beaver, dams were untended and began to fail. Erosion... on a grand scale over a vast area began and went unchecked. Environment, society, economy and worldviews changed on a global scale.

With the erosion, over time, animals, fish, birds, and people readjusted; but this would not always be the case.

The Saskatchewan River makes its way down from the Rocky Mountains, across the great plains and into Manitoba, emptying into *Kitchi Sakihkan* WINIPAKUWAM (Lake Winnipeg). By the time it got to Opaskwayak the river was wide, fast, and filled with silt. We, who lived on the banks of this great river spent hours and hours swimming, fishing, canoeing, playing in or by it… usually doing these pastimes unsupervised and on the *kimooch* (on the sly). We were told NEVER to swim in the river because it was dangerous. *Kitchi kinipik* (big snake) swam there and would grab unsuspecting children. Of course we never listened. We were untouchable. Impervious to such dangers… invincible. We never understood about strong undercurrents and the dangers that were possible.

We went so far in trying to hide our guilty pastime that we came up with a code word for going swimming. Early summer mornings would find us going around to each other's houses and when asked what we were up to or where we were going, it was either *sakak* (the bush) or *bah sil ah bah*, our code word for swimming. Don't know if we ever fooled anyone with this secret code because when we would come home later in the day we would be asked, "what were you doing? Were you swimming in the river?" To which we would reply "NO! We NEVER swim in the river!" At this point one of our caring relatives would take a damp cloth and wipe our forehead, face or hair and look at the cloth. It always had a smudge of fine dirt on it from the silt in the river! So much for swimming on the kimooch! But we always tried…

The Saskatchewan River was central in the lives of my family. My father's father, my grandfather, lived on the land and kept his family fed, clothed, and housed with the blessings from the Creator and the gifts of the land. We took our living from the bounty of this river and all the resources that came along with and surrounded it. The fish of all varieties: suckers, Pickerel, Northern Pike (jack fish), Burbot (mariah), Whitefish, Goldeye, and Sturgeon were provided by the river and its tributaries. Animals, berries, birds, bird eggs, plants, and medicines were all gathered from this plentiful environment. These resources were available for all who knew how to harvest them and all needs were met by the abundance of the river, swamps, lakes, and forest.

I remember hearing stories of Barrier, Big Bend, Raven's Nest, Big Eddy, Saskram, Cumberland House, Moose Lake, and the Summerberry. These and so many other places connected the people to the land and the land to the spirit of the people. The memories roll off the mind and into the realities of the people. These places were "home," not a square or rectangular house somewhere, but a living, breathing, giving and taking entity... *pimatisiwin*. In the language of the Ininew, this word, *pimatisiwin* can be translated as life, but it means so much more. This was the connection... to life... to the ancestors... to the future. Literally this word could translate as... moving forward... with the connection to ALL your female relatives... with confidence. The root word is *mitisi* or *mitisiyapi* (umbilicus or umbilicul cord).

As the people travelled the land they became intimate with it and all its resources. Their experiences interwove with the trees, rivers, swamps, rapids, portages, seasonal camps, and hills. There were laughing memories and hard memories... yet all set on familiar ground. One could predict known patterns and shared experiences. Experience and clear memory was the key.

This is where so and so was born... this is where so and so is buried... this is where we camped that time we saw the comet in the sky... this is the place you learned to track your first moose... this is where the sturgeon spawned and fish, eggs, oil, and bone were plentiful. This is where the eagles gathered before they went south... this is where *wikis* (pronounced wee guess), the all-purpose medicine was good... this is where you learned to be a man... this is where you learned to be a woman... this is where you learned to be a human... this is where... this is where... this is where...

Things changed in the late '50s and early '60s with the era of hydroelectricity.

The world conspired against my family and my people. Hydro development forced my grandfathers, grandmothers, father, mother, aunts, uncles, cousins, brothers, and sisters from the land they loved.

Canada decided we, as a people, were human and we were recognized as citizens… albeit marginal citizens whose labour could be exploited and whose rights could be trampled on when circumstance needed and profit demanded.

With the displacement of thousands of families throughout northern Manitoba to small towns like The Pas, or being relocated to new village sites, things were never the same. This displacement was the direct cause of flooding and unstable, unpredictable water levels due to the dams on the Saskatchewan River at Squaw Rapids and the building of the Grand Rapids Dam, as well as the Churchill River Diversion Project.

Keep in mind, this example is just one river… there were many hydro projects going on throughout Canada… throughout the world.

Livelihoods were destroyed along with families and whole communities being displaced… relocated. Generations that depended on the fish, fur-bearing animals, and migratory birds to make a living and have a sense of belonging could no longer depend on the natural patterns of their environment. Forced… torn from the land. Self-sufficient people who had lived off the land found themselves living near towns that resented them for their intrusion and blamed them for their confusion. "Stupid Indians"… "half-breeds"… "Why don't they go back to the bush?"

It seemed the only place that tolerated them were the bars and this is where a lot went.

This is the world I was introduced to and this is the world I grew up in.

My younger days were filled with loud violent drinking parties. I would wake up on any given morning and step over passed out prone bodies lying about the floor. If it was a weekday I would look around for any scraps left over from the evening's party for breakfast or go stop at my aunt Mary's on my way to school. She would always have something for me.

If it was a weekend I would scour the house and gather all the empty beer bottles and put them back into their cases. I would then pile them on my old Radio Flyer red wagon, tie them all down and pull them to the local beer vendor where I would return them for refund. Sometimes I would have ten cases of 2-4s (24 bottled case of beer) to return… $$$.

From there I would meet up with some of my friends and relatives and we would plan the weekend. Go to a Saturday afternoon matinee at the local theatre and then go fishing, swimming, hunting, or exploring depending on the weather.

This wasn't always our lot. There were times we could still remember when we were with our families… all together…all included… all important… all cared for… all with a purpose… all with food in our mouths and happiness in our futures…. I remember…

I remember tents in spring, summer, fall, and winter. I remember small log cabins with howling winds. I remember the smell of weasels and muskrat, of rabbit and beaver, of duck and goose, of fish and bannock. I remember walking down calm, silent trails in the winter when snow blanketed the trees. In that stillness you could hear the snow squeak. I remember the feeling that all was right with the world and a sense of utter calm and peacefulness. I remember laughing, fishing, snaring rabbits and weasels, skinning and scraping, stretching furs onto wooden frames.

I remember *Wawataywin*, the Northern Lights, static energy in the air and the smell of ozone… like after a lightning strike. I remember testing limits… whistling at the *Cibiyuk* (departed spirits) just to see what would happen. As children, we were told never to whistle at the Northern Lights or they would come down and get us! *Cibiyuk Nimintowin* was what the Northern Lights were called when they were in full color and swaying across the sky. The Spirits were dancing and they would come and get you to join them if you disturbed them. Of course we had to see what would happen. Looking back I see the reasons why we were told this. It was to keep us near the camp. It was winter and the only place to get a good view of the Cibiyuk was in an open area which for us was a frozen river or lake. It being winter and dark there were dangers to be avoided. Unsafe ice conditions under the snow, hungry animals hunting in the dark. I can imagine someone out in the darkness, in an open area, making all kinds of noise. Mikan, the wolf, could hear this too. Thus we were warned.

I remember being in awe of the night sky... staring up... wondering...

I remember I wasn't poor or hungry... in need... uneducated... unloved and unprotected... then I remember when all this changed.

In the late 1800s they began to send the children to residential school and they were changed. In the First and Second World Wars the men went away and those that returned were changed. In the 1950s there was the Korean War. The men went and returned and were changed. In the 1960s they welcomed the Indians into the bars and the world changed. In the 1960s they dammed the rivers of northern Manitoba and the world changed.

In the words of the great Canadian poet Gord Downie, "it would seem to me... I remember every fuckin' thing I know"...

This is where we of my generation were placed... this is where every generation has been placed, in their own time and space...

We all need a starting point. A place we remember. A space in time...

There is a place in my mind, your mind, our minds where ideas are formed, dreams are born, integrity lives and spirit is allowed to shine. No matter what happens, these places are always there. These ones determine where we will travel in our visit to this place we call *Aski* (Earth).

This is the place that others touch and connections are made. *Achakuk*... Spirits... shine. You are allowed to see them and you allow yourself to be seen. Life-guiding memories are born here.

When one talks of journeys, friendships, relationships, and ceremony these are the places we come… to pray, laugh, heal, cry, sing, love, and grow. From here we gather strength, courage, wisdom, understanding, clarity, kindness, compassion, and as my people say *mino pimatisiwin* (the good life).

This is what my grandparents (my mothers' parents and fathers' parents—four people) left me. It was here my grandparents allowed me to be me and in doing so left a beacon of hope that guided me through the years of self-loathing, self-deception, and self-doubt. If they never left this for me I would have never found my way through the cold, dark madness and hate that chewed up my spirit and spit me out… I guess I tasted bad!

My mothers' parents, Bella and Edward Ross and my fathers' parents, Carrie and Alfred Buck were my models of integrity, spirit, possibility, and kindness. These are the footsteps that I choose to follow. I came to see that these people were humans… a majority of the time very good and peaceful and at others (depending on the situation) very angry and hurtful. We are what we choose to be and for a lot of us, we are what situations dictate… demand and deplore.

As I walk this path I have come to realize that I learn and have learned by example. I had a very hard time in school because people were always telling me what to do, how to think, how to live, what I should feel and who I should be. Very few people that I could see, lived what they professed was the way to live. The dominant society even told me what I should believe.

I remember I was sent to Sunday School when I was about five years of age. I did not want to go; I was forced to go by my well-meaning aunt and French-Canadian Catholic uncle. I remember listening to the Sunday School teachers talking about Hell and all those who would go to Hell. I asked if my relatives were in Hell because they lived and died on the land. I was told they were in Hell because they were not buried on sacred ground. I said they were buried on scared ground because our Mother Earth is sacred. They told me not to talk so crazy or be so superstitious. I told them I would like to go to Hell because this is where my ancestors were. I walked out of that place and never went back. But I knew the church held sway over my people.

I guess this conflict of faith and acquisition of wealth is a basis for the tension felt by the "Indians & Cowboys" of the immediate area which manifests itself in guilt, hate, and violence.

The animosity felt between the local townspeople of The Pas and the people of Opaskwayak was palpable. There is a history to this.

When Europeans arrived in the area in the mid 1700s, my ancestors occupied the area around Devon Island on the south shore of the Saskatchewan River, near the Opaskwayak River. Opaskwayak was an important camp for my people in the many camps which made them intimate with the environment they found themselves in.

French traders established a fort in the area and the fur trade flourished.

The site was very sacred to my ancestors and was seen as a healing place. Very near Devon Island on the south bank of the Saskatchewan River, a meteor had impacted and was partially visible on the bank. The people would come from miles around to hold ceremony and offer up *wipinasoon*, prayer flags and tobacco for healing, safe journeys, and good health. The sacred rock was seen as an offering from Creator. The energy emanating from that sacred rock made people feel at peace. It was a holy place.

In the spring and fall months, the sacred "Goose Dance" was held paying homage to the sacred gifts *Niska* (the goose) gave to the people.

Things always seem to fall into place. Niska has had an impact on my life.

As a sun dance chief I had dreams about Niska falling from the sky. I was told this dream meant we as sun dancers could use the goose bones as a substitute for eagle bones for our *pipitacikunuk* (eagle whistles). With the resurgence of the sun dance among our people, Niska the goose would help *Mikisew*, the eagle, so that so many eagles would not be killed. I read somewhere it is now estimated there are as many as 2,500 sun dances held throughout North America during the summer months. That is a lot of eagle bones if at each sun dance there are anywhere from ten to one hundred and fifty dancers. Thus dreams, history, and reality intermingle and we are blessed, given direction and healing.

Another aspect is that I was given a mythology about Niska flying in the milky way… which my people call *Niska Meskinaw* or *Nipinesisuk Meskinaw*… the goose's path or the summer birds' path.

The Niska mythology goes like this… Situated inside the Milky Way is a group of stars commonly referred to as the Northern Cross. Roman Greek mythology call this group of stars Cygnus, the Swan.

My people refer to this group of stars as Niska, the Goose. The story is that whenever these migratory birds were flying, hunters would go out and hunt them. At times the hunters would stay out all day long hunting these *Pinesisuk* (birds). As it began to get dark, the hunters would return to their camps. Sometimes as they were travelling back to camp in the near night sky, they would hear the whistle of the goose or duck wings beating the air. At times they would look up and see the silhouette of the geese or ducks moving against the backdrop of the Milky Way. Thus the stars Deneb and Albireo were identified as the tail and head of the goose and the stars of the Northern Cross are called Niska, the Goose. The path Niska flies in (Milky Way) is referred to as Niska Meskinaw, the goose's path or Nipinesiwuk Meskinaw, summer birds' path, for this is the path the migratory birds use to fly south and return north.

**NISKA MESKINAW (the Goose's Path)/
NIPINESIWUK MESKINAW (Summer Birds' Path)**

These birds were a very important food
and medicinal source for the people.

When the Anglican Church arrived in 1840, the first order of business was to get rid of the false idol the Indians prayed to. The missionaries tried to blow up the meteor and failed so they did the next best thing—they rolled the meteor into the river and then built their church on the very site the *Manito Assisni* (Spirit Rock) had sat. The peoples' spirits were wounded.

In September, 1876, Treaty 5 was signed between the Indians of The Pas and the Crown. The people were relegated to a reserve and the whims of an "indian agent." During this time the missionaries and indian agent rounded up all sacred items the Indians possessed which connected them to their old ways and burned them or threw them into the Saskatchewan River. Sacred Pipes, rattles, drums, whistles, feathers and fans, painted hides; everything that held any spiritual significance was destroyed. Once again, the peoples' spirit was wounded.

The next assault to the spirit came on August 4, 1906 when all the natives were rounded up and the head men were forced to sign (at gun point) a deed to surrender land occupied from time immemorial by the people. They were then marched (again at gun point) across the new iron bridge to the north shore of the Saskatchewan River where the present day Opaskwayak Cree Nation now sits.

It should be noted here that these assaults on the spirit, identity, confidence, and integrity of the people of Opaskwayak is only one instance in all of what is now Canada and North America. These genocidal policies were and are still happening throughout the AMERICAS.

Now with Treaty 5 signed, the indian agents of the federal government took over the affairs of the people. Indians were confined to the reserve, forbidden to leave the reserve or communicate with others outside their immediate reserve. At one point people could not even meet in groups of five or more without the written consent of an indian agent. Everything that they did or wanted to do had to be first presented to the indian agent for approval. They could not so much as gather firewood without consent. The people flocked to the churches for comfort and hope.

A brief synopsis from one of our young scholars… Acak Nimitowin… Ivana Yellowback…

> The Northwest Mounted Police (later turned RCMP in 1920) was created in 1872 by Sir John A. Macdonald as an adaptation from the Royal Irish Constabulary model.
>
> The purpose was to have a force put forward to diminish "Indian Wars" like those occurring within the United States at the time, to further extinguish and colonize Indigenous people's rights, lands and territories to further push the Dominion of

Canada into the west and secure "investment frontiers" within Manitoba, Saskatchewan and Alberta. It was stated that "once they had serviced these duties they would simply disappear."

However due to the pushback from Indigenous nations (with leaders such as Chief Poundmaker, Chief Big Bear, Louis Riel, etc.) they then created the Stoney Mountain penitentiary (within Manitoba) to harbour and jail Indigenous leaders trying to cause "rebellions" or remain practicing their sacred ceremonies and Sun dances, all of which were deemed as "political acts of threats."

Following the implementation of the Indian Act they then further perpetrated the "Pass System" whereas no Indian(s) were able to leave reservations without noted jurisdiction from a non-Indigenous Indian Agent(s). Any Indian(s) seized off a reservation without one was deemed a "political threat" and to be "conspiring towards the Government" and jailed.

They also implemented the forced removal of young Indigenous children and youth during the Residential School era, and were further placed onto reservations to police and enforce colonial laws on Indigenous communities.

So, when individuals (especially emistikosewak) create a point to complain about young Indigenous men and women not wanting to respect or adhere to the police/RCMP, know that these struggles and feelings of anger/hatred have been active since the mid-1800's.

The implementation of colonial laws and euroaggressive-epistemology have been active for over 500 years; just like the "symptoms" of resistance, inter-generational systemic-racism and "internalized-oppression" has been active for over 500 years.

It is interesting to note here that this system imposed on my people was studied by certain South African interests where it was exported and imposed on the Indigenous people of South Africa... renamed Apartheid...

As well it can be noted that this type of social, emotional, racial, and cultural genocide took place all over North America. What has happened to my people happened, and still is happening, to Indigenous people *all* over the globe.

All seemed to go the townspeople's way until the early 1960s. At this time the chief of Opaskwayak, Chief Cornelius Bignell, began to look into how his people could make a better living and in doing so found legal means to begin to reestablish governance over The Pas Band lands. Up to that point in time the town of The Pas pretty much expanded and expropriated native land as they saw fit. Now they were in fear of losing what they had taken.

It was also during this time that a booming pulp and paper industry sprang up in the area and workers flooded in from all over the country. It was the "wild west" and Indians were again losing out. This was also around the time that Helen Betty Osborne was abducted off the streets of The Pas and murdered. Helen Betty Osborne was not the only Indian found dead during this time though. There were huge confrontations between townspeople and Indians on the streets of The Pas. Three major fights involving 200–300 people took place.

I remember one fight that involved numerous people having a pitched battle on a street in The Pas as RCMP observed the situation. The confrontation began with some French Canadian pulp mill workers mistreating some of the women from Opaskwayak Cree Nation. Physical abuse was taking place and some of the men of the reserve objected to this. Fighting began between the two groups. Soon sides were divided along racial lines… indians vs whites. This feud spread out from an initial confrontation to a three day running battle with both sides picking out a headquarters or base of operations. From these HQs raids were carried out into enemy territory.

All this came to a head on a warm Saturday night in June when we were planning our strategy in our HQ and someone came in and said, "They are coming!!" Some of us went outside and confronted the initial assault. We noticed the RCMP had cordoned off the street and were watching the "show."

The "bad guys" came charging down the street en masse. We confronted them until we were overwhelmed by numbers and retreated back to our HQ.

We sat inside our HQ while the building was peppered with all kinds of projectiles and every window and door was smashed. Meanwhile the RCMP did not do a thing.

It was decided we would charge out into the street and put an end to this one way or another.

We came storming out onto the debris littered street to confront the enemy. They threw everything they could at us. We came on. The enemy saw our determination… faltered and then fled. We chased them to the end of the block.

At this point we were confronted by a row of RCMP officers dressed in riot gear. They stood spread out across the street, facing us with riot guns (shotguns) in hand.

They told us to stop where we were and return to the hotel barroom.

It was then I happened to look at the roof tops of some surrounding buildings and saw police with rifles standing there looking down on us!

We were herded back into the hotel and told to stay where we were.

The RCMP let the "other side" go without any ramifications whatsoever.

The RCMP then arranged for buses to pull up at the back door of the hotel and we were loaded onto the buses and driven to the reserve. It looked as if we were being smuggled out of town, "for our own protection."

At this point we were told not to return to town until further notice. Northbound traffic out of town was allowed to flow while southbound traffic containing Indians was sent back to the reserve. This lasted for the rest of the weekend.

To make sure we knew our place, the RCMP stationed guards on the reserve side of the bridge so we could not get back into town for revenge.

The following week things slowly got back to normal but still cars were being searched. One of our party, Cecil Bignell, decided he would jump across the ice flows to get across the river since the RCMP were guarding the bridge. He took with him a sawed-off shot gun. The RCMP were searching cars, and if no weapons were found, letting them go into town. Back to the local bars we went with tension hanging in the air like a dense fog. We met up with Cecil and the shotgun was passed around. It was decided that we would split up and scout out the local bars. Another of our party, Patrick Bignell, got arrested for taking the same sawed off shot gun which he pointed at an undercover officer outside another local bar when the officer pulled a gun on him without identifying who he was. Patrick got sent off to prison for a while.

We did not realize it at the time but we (Indians who talked about the American Indian Movement …AIM) were under surveillance by the RCMP. I remember this one time, not too long after Patrick got picked up, we were sitting in a local bar and I was rambling on about how we should invite AIM up to The Pas and maybe get a hold of some explosives and blow up the local bridge! Of course this was bravado but nonetheless this information was heard by someone who informed the local RCMP and I was dragged out of the local bar and taken in for questioning!!

This was the backdrop for Helen Betty Osborne… the wild west… cowboys and Indians…

Just on a side note: I had chance to meet one of the founders of the American Indian Movement at a later date—Dennis Banks, Anishinabe of Leech Lake Reservation. He was very interested in the knowledge Indigenous people hold regarding *achakosuk*, the stars.

I HAVE LIVED FOUR LIVES...

My visit to this place began on a Monday, September 27, 1954 at the Saint Anthony's Hospital in The Pas, Manitoba. I was one of three children given to Louise and Alfred Buck for safe keeping. Peter and Rose were my older brother and sister.

Alfred Buck, my grandfather, Carrie Buck née Young, my grandmother, lived with their family and extended family at a camp on the Saskatchewan River, at a site known as Barrier. It was from here my grandfather and grandmother provided for their *nitootimuk*, relatives. It was at this place, my grandfather's camp, west of the town of The Pas, on the Saskatchewan River, that my sister Rose passed away from influenza when I was two. During this time our family lost two other relatives, my uncle Walter and his wife lost two children as well. This combined with what was in the planning stages for hydro electricity production, wrenched my family from the land.

After this point, it seemed the world picked us up and dumped us in another time continuum.

One minute my grandfather and his family were self-sufficient people living off the land, and the next we were deposited into the urban reality of marginal living on the edges of the town of The Pas, Manitoba.

My grandfather taught his sons all they needed to know to live on the land. He taught them to read the seasonal changes, animal tracks and habits, how and where to set nets, snares, and traps. When to go harvest eggs, medicines, willows and the various wood used for smoking meat, and making racks and pelt stretchers.

It seemed one morning all this knowledge was of no use anymore. Everything was flooded. The world was no longer the same. My grandfather, grandmother, father, mother, and uncle, aunts and cousins were displaced. All they knew had no meaning!

This happened to families all over the north and along the Saskatchewan River from the Saskatchewan border in the west to the Grand Rapids in the east. Livelihoods destroyed. Lives thrown into the turbines of hydroelectric progress.

My father slipped into depression with all the sudden changes and not being able to make a living for his family, began to find the numbness of alcohol.

My mother, not being able to keep things together and depending on her sisters for the care of my brother and I, followed my father into the numbness.

This is when the family I knew was thrown to the wind. Peter, my older brother, was sent to live with my father's parents and I was sent to live with my mother's sister and her husband.

Soon after this my parents split up. My mother began living with another man and had five other children.

My father also began living with another woman and they had two children. Thus I had nine siblings, none of who I grew up with, some I didn't even know were alive.

I remember going to visit with my mother and her new family. I remember playing with my siblings but the only name I could recall is *Mah Nah Tis* (Ugly), one of my younger little sisters.

She was called this as a tease. She would always say… "look at me… I am pretty." Whereupon someone would say, AAWUS MAHNAHTIS! and jokingly laugh.

I was staying in the town of The Pas at the time with my mother's sister, auntie Jean, and her husband Art, my uncle. He was very strict and they both used to drink and have huge house parties all the time. I usually was left to my own means.

On weekends and sometimes when I would skip school I would go see my mother and siblings. I used to sneak away and walk three miles to visit them. I guess I was 8 or 9 at the time and I remember walking across The Pas bridge, sometimes on the narrow roadway and other times on the railroad trestle. It was a scary trip but I remember doing it in sunshine and in driving rain.

In the winter the journey to visit was shorter. I just had to walk across the frozen Saskatchewan River. The winter made the trips to visit less frequent also, because every Saturday my siblings and cousins would walk by my place on their way to the Saturday afternoon movies.

I looked forward to these Saturday meetings and would gather up all the beer bottles laying around my house from the Friday night drinking parties and go sell them at the local vendor to get enough money to attend the afternoon matinee. Sometimes when I woke up on Saturday morning the party at my place was still in full swing and I could get extra funds for running to the store for cigarettes and soda pop mix for the partiers.

Then one bright summer's day when I was on one of my many visits to see my mother and younger siblings before making my way to see my older brother, grandparents, and cousins, once again the world changed. As I made my way through the bush (a shortcut to grannie's house so to speak) I encountered a wolf sitting on the path, blocking my way. I was totally in fear! I slowly backtracked and tried to go the longer route, down the rough bush road and once again there was the wolf sitting on the road blocking my progress. Again I tried another route. This time I tried to walk along the river's shoreline but again there was the wolf!

I once again backtracked and sat down to decide what to do.

As I sat by the worn path that was the shortcut to grannie's house (and my mother's as well), I heard a car go by, driving away from the small community where my relatives stayed.

I snuck a peek from my hidden vantage point and saw this old car driving slowly down the rough road. I noticed small shadows of heads in the rear window.

I got up and thought, that car should scare the wolf away. I once again set out down the worn bush path to see my mother and younger siblings. This time the wolf was gone! I slowly shuffled along the path keeping a wary eye out for the big bad wolf but it never showed itself. I was in the clear!

When I arrived at my mother's place there was no one there. Everyone was gone. Not a soul around other than the camp dogs. So I continued on to see my older brother, cousins and grandparents and to tell them the story of the wolf.

I would learn later that the wolf, *mikan*, was one of my guides and helpers. *Pahwakan.*

I never saw Mah Nah Tis again for a period of forty years. My siblings were victims of what has become known as the Sixties Scoop, a government sanctioned apprehension of native children who were adopted and shipped off to everywhere but home. At the time my sister would have been 3 or 4 years old.

My mother was out doing casual labor during this time. The children were left with an old couple for safe keeping. We called this couple grandma and grandpa Badger since this was their last name. Apparently whoever came to get the children knew that my mother would not be there at the time. All siblings were taken and my mother never saw them again.

Thinking on this chain of events now it would seem Mikan saved me from also being included as one of the statistics of the Sixties Scoop. If I was not delayed on my journey to visit my siblings that morning, I could have been with my siblings and been stolen as they were stolen.

After this time my mother left town and went to the city of Winnipeg where she drank herself to death. She died on the streets of that city like so many of my people.

I stayed with my aunt and uncle from age five until I left at the age of thirteen. During this time I grew up surrounded by alcohol abuse and violence. There was always drinking parties going on, sometimes lasting for days on end.

We moved from place to place, from rented farm plot to the town of The Pas and from place to place within the town. The parties always followed along with the violence. At these times I remember eating macaroni and tomato soup, sometimes with wieners or sausages thrown in for flavour, a lot of times not. I remember being sick of eating macaroni and tomato soup. I remember when I first heard of MTS (Manitoba Telephone System) I right away thought, Oh no, not Macaroni Tomato Soup again!

It seemed I ate better when I went to the bush on the weekends and summer holidays during my youth. Friends and cousins would gather every morning and head out to the bush to swim, fish, catch frogs, snare rabbits, shoot prairie chickens with sling shots, and eat berries.

We spent countless hours out there until it got dark and we had to head back home.

Some of us regretted having to go back home because we never knew what we were going to find or what imagined trouble our guardians would think we were up to. Uncertainty, confusion, anger, hurt and loneliness filled our nights at "home."

Always a soothing pastime, hunting was an excellent way to forget my immediate situation. All the bullshit, racism, and pain could be forgotten with the thrill of the hunt. I remember one time I was hunting alone and was chasing game through the forest. I was so intent on catching up to my quarry that I was not paying attention to my surroundings. I came to a stop, at the edge of a small clearing in the forest to catch my breath, and slowly realized

I did not know where I was! I WAS LOST! A cold chill ran down my back and an immense panic washed over my body. I started shaking and sweating and got a very strong urge to start running and screaming through the forest... PANIC!

I then remembered a stupid saying we had... "don't panic... have some bannock." So I sat down, got out my small lunch and had some bannock. As I did this I realized I was not lost. I was east of the community and the river ran in an east west direction and should be south of my position. I looked at the sun and stuck two pairs of twigs into the ground in a line along my shadow, each pair about three feet apart, with about three inches separating each pair of sticks. I then sat down and waited for about 15 minutes. I watched the shadow the twigs cast move. From this I knew where east and west were, therefore which way I needed to travel in order to get back to either the community or the river. I found my way.

I know of one of my friends that did not find their way out.

The story goes like this.

One day a couple of us were returning from a hunt and we noticed our buddy Dan jogging up the trail all by himself. Dan liked to stay in shape and he had decided to run up this trail for a few kilometers, have a rest, turn round and jog back. We told him, "have a good run... see you later," and continued on our way. That evening we saw Dan's family going around asking if anyone had seen Dan. We mentioned the last time we had seen him he was jogging up the trail heading east. We were told by his family Dan had not been seen for a while. It seemed we were the last people to see Dan. He was lost and no one knew where he was.

Dan was found three days later running through the bush, still screaming, totally out of his mind. He got sent away down south for shock treatment we were told. I remember when I saw Dan again a couple of years later; I had just returned from one of my many journeys west and a couple cousins and I were in a taxi heading for town. As we were driving down the gravel road, I noticed someone standing by the side of the road. He was standing there doing a "Heil der Fuhrer" salute, holding a small black comb under his nose.

My cousins said "Yes, that is Dan. After Dan was sent home from his shock treatment he has not been the same. Dan has not recovered from his time in the bush and the shock treatment."

What had happened to Dan was that after we had seen him jogging down the trail that day, he decided he would sit down and have a "joint" he just happened to have with him. This he did and after a time he decided to jog back home. Only thing was he had forgotten which way that was! He had lost his way. At this point panic overtook him and he ran into the bush, screaming. Poor old Dan. He did not have any bannock!

I stopped visiting my relatives across the river on a regular basis after my siblings vanished.

At the time I must have been 8 or 9 years of age. I withdrew into my own little world. I began spending every scrap of money I could get my hands on buying model airplane kits. The house we were staying in at the time in The Pas, Manitoba, had an old shack behind it and this served as my base of operations. Here I built, painted, and hung up all my model World War II airplanes.

I sold empty beer bottles gathered up from Friday night drinking parties. There were many parties at my house so there was no shortage of "empties." Every Saturday morning would find me roaming over and around passed out bodies who over indulged in the festivities, collecting beer bottles. There were all kinds of things I would find in these bottles: cigarette butts and beer, KFC bones and beer, condoms and beer. I guessed everything went with beer (I was to find out later on in life everything went with and HAD to accompany beer). The reason I knew what was in the bottles was because in order to sell them, the bottles had to be empty. If I did not empty them then when I packed, transported and unloaded them, they would spill and get everything soaking and smelling like stale beer and whatever.

There was a lot of drinking so I had a lot of empties to cart off and therefore money to be spent… models and accessories to be acquired.

World War II Model Planes… RAF Spitfires, Japanese Zeros, Curtiss P-40 Warhawks, Hellcats, Corsairs, Lancaster Bombers, de Havilland Mosquitoes, B-17 Flying Fortresses. The list went on and on and on. I even began building planes of the Korean War. During this period I must have had approximately 300 models… all painted, detailed, and hung all over my sanctum. Hours and hours of patient work were spent finishing every detail. Paint bottles, paint brushes, glue, knives, and empty model boxes littered my refuge from the world.

I guess it was a healing therapy for me from the uncertainty of the world. A place where I could lick my wounds. This period lasted about two years. During this time a number of things happened that changed my point of view. Everything went to shit.

This was also the time my uncle and grandfather hung themselves. My mother's brother and father. This was also the time my aunt finally had had enough of the beatings she was given, it seemed on a regular basis, and left for the big city. There I was, it seemed, being left alone again. My mother, my father, my siblings, my uncle, my grandfather and now my aunt; who else was left? It seemed I was left to my own resources.

Had to change my way of thinking. Had to go into survival mode. Had to look after myself. Had to hurt before I was hurt. Had to destroy before I was destroyed. Had to take before I was taken. Had to put me first before all else. Had to leave before I was left.

I burned all my models and began hanging around with others who were as angry as me. Started drinking and stealing, breaking into stores, houses, automobiles, railway cars and beating on people.

This was the beginning of the long dark night. A night that would see me dead to many people and a night where I would see myself dead.

So began the long cold journey into self-pity, hate, anger, confusion, loathing, sickness, and loneliness.

I was eleven, an alcoholic who had never drank alcohol. I had all the symptoms but needed the "juice" to complete the transformation.

The very first time I drank alcohol I drank until I blacked out because this was the only way I had seen alcohol consumed. Our little gang used to hang around on the outer edges of town at the ruins of the old lumber mill where cement structures still remained; a good hiding place, meeting place and planning place, away from the watchful eyes and inquiring minds. One day we were there planning a raid on a local social hall. A social was a place adults would go to have a dance and sell alcohol to raise funds for other activities such as weddings. It was a good bet there was money or other valuables there, especially after a function.

While we were planning the great raid we heard a car pull up and everyone hid. We thought it was a parent hunting someone down. As it turned out it was a taxi driver. He was in the process of hiding his bootleg wine he was planning to sell the following Sunday, or as we called it, "Hang-Over Day."

We hid and watched him hide his stash by the river bank. He then got into his taxi and drove away. We went down to the stash and found four cases of Jordan Branvin wine! This was the cheapest wine someone could buy to do the job. The party was on!

There were six of us and three of us drank until we could not walk... or remember. The next thing I realize after drinking the first bottle of gawd-awful, throat-ripping wine, is lying in my bed, crying and crying.

What had happened. I found out later, was that we raised so much noise and commotion that a passerby called the Royal

Canadian Mounted Police on us. When they arrived we were placed into police cars and transported home since this was the first offence for us and we were from the local area.

Apparently, I received a beating for being bought home by the RCMP in an inebriated state thereby causing the eye of the law to throw its focus on my uncle. He was not amused that we had squandered the ill-gotten gains on ourselves. We/I should have come home and showed him what I/we had found so he could have taken the bootleg alcohol and made some money.

My uncle began to take notice of my friends and our night-time activities and pretty soon he was helping us plan our break-ins and transport our goods. He also began to be the one to buy us alcohol and we would party at my place. At this stage of the game I thought life was good.

This went on for a couple of years and slowly affected my schooling. In grade seven, I was twelve, when a friend and I got into an altercation with a teacher in the classroom and we were expelled from junior high school. Our principal was furious at us. One thing I remember about our principal was that he liked to wear smart suits. He always wore nice looking suits, no matter the weather or occasion. The other thing I remember was that he had two missing fingers on his left hand. So anyway there we were sitting in his office being yelled at by him, "you two are going nowhere in life! You will be just like your parents...drunks and on accounts! You will probably be in jail by the time you are teenagers! Why should we waste our time with you? I can give you five good reasons why you are going to be expelled today!" With this last statement he held up his left hand with the two missing fingers...five good reasons! That was it! We could not contain ourselves any longer. We both bust out laughing hysterically... holding up our hands with two fingers bent and crying, "five good reasons!"... Thus our schooling was done.

I was now free to dedicate myself to mischief, mayhem, and marauding. We got so good at it that the white kids of the town started calling us "The Marauders."

A group of us would gather as night began to arrive and plan our activities. One of these times we broke into a storage shed and found two large camp tents. We thought… these would make awesome party facilities, so we took them. We set them up in a bush near the river and then went about planning for a big party.

It was early June and we somehow managed to steal twenty cases (2-4s) of beer. We spread the word and our PARTY was underway. To supply snacks we broke into a food store and "went shopping"—took what we wanted. The PARTY WAS ON! At one point I counted about thirty youth partying up a storm.

The last thing I remember before I blacked out on alcohol and drugs was seeing all kinds of youth going wild… dancing around a roaring bonfire in the bush as sparks floated up into the night sky.

When I came to I found myself in a little jail cell on the cement floor… shivering and very sick from a pounding hangover headache. Apparently what had happened was we made so much noise at our party that someone called the RCMP and we were raided.

Upon questioning the majority of party-goers, our core group was singled out and we were arrested for a long list of offences. For a number of us, who were known by the police as being juvenile delinquents, we were charged and eventually sent to detention in Winnipeg.

This was my first sight of the city… through bars. Summer in the city was spent gazing at Memorial Park through bars from the Vaughn Street Detention Centre in Winnipeg. We spent five months there and were eventually released and returned back to The Pas. Free… free at last!

It was about this time I decided that I had had it with home life and all the rules and contradictions that came along with it. One Sunday night in early fall… I think it was late September, I was sent to the store for some cigarettes. I remember walking to the store feeling really lonely and depressed, thinking no one would miss me if I decided to disappear. So, I just kept on walking. I walked into the darkness and never looked back.

I walked into another world…

||

I walked into the loneliness of a cold harsh world… but it seemed I was ready. I lived in used car lots… couch surfed… stayed at parties on weekends and hitch-hiked to other locales. I survived by being a thief, liar, burglar, shadow, drug dealer, stolen goods fencer, and quick of wit… resourceful… unreachable… untouchable. At times I was a coward to my enemies, hero to my peers, and a scourge to the law-abiding citizens wherever I landed.

Living on the street is an empty, cold, lonely affair. Luckily there were places I could go to fill some of the void in my heart and stomach. Still, a strong cold wind blew through the giant hole in my spirit. I was lost…

My time spent on the streets was a series of misadventures where drugs and alcohol were central to my existence. I lived for these items. My world revolved around these items.

During this span of time I was run over, beaten with two-by-fours, stabbed, shot at, fell through the ice… twice, in car crashes, had an alcoholic seizure, and overdosed. Misadventure after misadventure and always alcohol and drugs were integral to my interactions with the world. No alcohol or drugs, no Wilfred Buck.

I was run over… twice.

The first time I got run over I was bumper sliding in a back lane of The Pas, Manitoba. I must have been eleven then. I snuck up behind this taxi, as it was driving slowly down a back lane, and went for a little ride. Apparently the taxi was looking for a fare in the area and as I was sliding along, the cab came to a very fast stop. I slide under the rear of the taxi. Just then the cab driver began to back up and I got my legs run over. I was laying there screaming bloody murder! The cab driver got out scared and screaming!

I was lucky! Nothing was broken and I was still alive. I walked away cursing the so and so cab driver! Fucker was trying to kill me!

The next incident involved a semi-trailer and happened on the outskirts of Dauphin, Manitoba.

During a hard drinking spree a number of us ended up in Dauphin at a hockey tournament. I remember riding in the back seat of a car on the way to... somewhere...

There were four of us in the back seat and I do not know how I ended up with these particular people because normally we were enemies back home. Anyway there I was riding with them. One of them mentioned he had pulled a knife on a friend of mine and they all laughed. I laughed too... as I punched the guy in the face! The fight was on! During the mayhem, someone opened the door of the vehicle and I was thrown out onto the highway!

I remember getting up off the pavement... staggering around and falling down again. I remember thinking... I will just lay here for a while... then I must have passed out! Again Death came and gave me a look over. Something watched over me.

The next thing I know I am being dragged by my hair into the back seat of an RCMP vehicle.

The RCMP Officer was screaming and swearing at me.... "You dumb son of a bitch! A semi just about ran you over! Swerved at the last second when he realized it was someone laying on the highway!" I remember screaming, "Police Brutality... you're doing this because I am Indian!"... not because I was a dumb son of a bitch!

We did an awesome amount of drinking and drugs. This was our life.

I remember one time our little gang somehow got into the local bar's vendor and stole about 200 cases of beer. We got some of our older friends with driver's licenses to borrow cars and then drive them to the back of the vendor where the suppliers loaded the beer into the hotel's cooler. Here two people went inside the cooler and passed the cases of beer out to others who loaded them into waiting cars and drove off to unload, then return for another load. The party was on!

To this day I do not know how we got away with this. Our small town had a reputation of hard drinking and lots of fighting. I guess there was so much drinking going on all over the place that our little party with our ill-gotten gains was never even noticed!

Numerous break and enters (B&Es) netted me the designation of criminal mastermind by the local RCMP and apparently I was my own crime wave!

When our little crew was not drinking, marauding, and raising hell, some of us spent our time making up songs and twisted sayings. We would make up our own lyrics to popular songs and sing them. One such creation has stuck with me through all these years and I do not know WHY. This particular song was sung to the tune of The Beatles' "I Saw Her Standing There." Looking back, we did this to entertain ourselves, but with all I know, I wouldn't do it now. It went like this:

Well she was just ninety three
With a touch of leprosy
And the way she looked
Was way beyond repair
So how could I dance with her mother?
When I saw her standing there…

We murdered a lot of songs in this way and had a good time doing it.

Some of the sayings we twisted around were things like "built like a shit brick house, useless as a lipless man at an ass-kissing contest, thick lips… float ships, you can't teach a dead dog new tricks" and the favorite… because of the shock value… "you can't keep a good man… nailed to a cross!"

I lived everywhere and anywhere I could. During this time I stayed with various friends and family on the reserve, in the town of The Pas and the Métis community of Umpherville Settlement (the U.S.). There are still some in our Opaskwayak Cree Nation community that remember our partying group we called the Dragons… the U.S. Dragons.

Another time, in a cold January, our crew got into the local liquor store and had acquired about 300 bottles of whiskey, gin, vodka, and tequila. The party was ON! This was about the same time we had shot a moose and we had it stored in the wood shack out back of our party house.

We drank, got sick, ate moose meat, drank, passed out, got sick, ate moose meat, drank, fought, got sick, ate moose meat, drank… and on and on it went.

On the last leg of this wild binge, I woke up in the dim light of the early morning or early evening, I could not tell which. I awoke to dim shadows moving around the dark house. I laid there and talked with a shadow that was sitting in a chair by the bed I was on. I talked for a long time and got no response. I finally asked for a cigarette and was handed one. I felt around for a light for my cigarette and found one by my bed. I sat up and lit the cigarette and as I did so I saw I was alone! There was no one in the room with me! I jumped up and ran to turn on the lights. I look around and no one there. I checked all the rooms. No one! I started to shake. I started to sweat. My heart was racing. I was scared!

I took off out of the house and headed for the nearest place I knew people would be. The neighbors. Turns out it was evening and as I told my hair-raising story I was told maybe I should stop drinking. I said, "YEAH… CHAAA! You Think?" I quit for three weeks and then took off from that crazy-ass rez, with all those crazy-ass people in it. They were NUTS!!

It was also during this time that I learned how to run away and that I was very good at it. I found this solved a lot of my problems… running away. I remember my mother running away from my dad and going to live with her sister and her husband, my aunt and uncle. I remember being left there with my aunt and uncle. I remember my mom leaving… saying goodbye… crying. I remember chasing her across a field as the rain fell. I remember crying… crying.

I remember my aunt running away from my uncle and again I was left behind. I remember running away from my uncle's place. After that I realized I could run away from anything I did not like or agree with.

I spent about eighteen years running and during that time I became intimately acquainted with every skid row in Western Canada. Living in used car lots, shelters, hostels, flop houses, abandoned shacks, and tents. I learned from others and survived.

At times I ran as far as the Pacific Ocean.

September 1970 found me in Vancouver B.C. I remember the time because this was when the album and the song "Paranoid" was first heard over the radio waves. The first time I heard the song "Paranoid" I was hitchhiking with a buddy through downtown Vancouver on our way to a party somewhere. I heard the song on the radio and asked my buddy, "who plays this song?" to which my buddy replied, "Black Savage." I said "COOL…Black Savage eh? This is a good song!" It was only a week later that I discovered the band was not Black Savage but rather Black Sabbath. Ozzy!

By this time I had discovered LSD, MDA, STP, and P O T, and along with alcohol my mind was gone…

It was also during one of the times I spent in Vancouver that I first heard of the American Indian Movement (AIM). I lived on the fringes of sanity and once in a while I would get an interest that did not involve drugs or alcohol. I began to hear about Indian ceremony… sweat lodges, sun dance. I had heard about these things before but it was not up until that point that these ideas started to sink in to my mixed up mind. Something began to grow.

One December I hitch-hiked back from Vancouver… a journey that took one week, from Vancouver, B.C. to The Pas. I left Vancouver walking in the rain and walked into the full force of winter in Canada on the Prairies. Outside of Regina, Saskatchewan one cold winter night, I was trying to catch a ride to Winnipeg when a blizzard blew up. IT WAS COLD! I remember shivering and running to whatever side of the highway a vehicle was coming. I did not care which way it was going; I only wanted some place out of the cold wind and snow to warm up. I ended up back in Regina for the night. My next stop was Brandon, Manitoba and then north to The Pas.

A couple of days later I arrived in a small town called Erickson, Manitoba, just on the southern edge of Riding Mountain National Park, on provincial highway 10. Once again a blizzard blew up. This time I knew I had nowhere to go. Small town Manitoba shuts down at 5 p.m. I was stranded in a blizzard on a dark highway. I knew I had to get out of the howling, cold wind. I stood there for as long as I could but no one was driving in this crazy weather. As I stood there I noticed an old weather-beaten shack out in the middle of this field off the highway. I decided this was where I would spend the night. I made my way to the shack and it was out of the wind… cold but out of the wind. The walls were wind dried boards in which at some places you could look outside through the cracks. There was minimum shelter, walls but no door… but beggars can't be choosers.

I laid out my sleeping bag and put on every bit of clothing I had. I took off my boots and placed them right beside my head, climbed into my bedroll and went to sleep. The wind howled and the snow flew.

During that long cold night as the wind whistled and moaned I had a strange dream. I dreamed of Sasquatch and how it materialized and came trudging through the snow. Then it was gone.

Woke up the next morning and the wind had stopped blowing. The sun was out and it looked like a good day. My boots were frozen to the floor! I decided to build a little fire to thaw out the boots as well as my chilled bones. When I finally got them on I packed up and went outside. As I left my refuge I noticed animal tracks around the shack where I had spent the night. I thought nothing of it and walked to the highway and found an open gas station. While I drank coffee and thawed out I listened to the local people talk about the blizzard and how cold it had been.

While they talked, a conservation officer entered the gas station and joined in the conversation. He said that during the blizzard the wolves were howling like crazy and he had found a half-eaten elk, "not too far from here," he said.

I straightened up and took notice! Were those wolf tracks I had seen outside my lodging? Cool! It never occurred to me that I could have frozen or been food for the wolves!

It also never occurred to me that *Mistapew… Sa bay…* Sasquatch, also had some connection to this chain of events. In dreams they come…

It was not until later on that my cousin Leonard told me a story about something that had happened to my *mooshoom* (grandpa) while the family still lived off the land. It was told that one day while my grandfather was out checking his traps he was making a portage and heard a strange sound coming from a large tree just off the portage trail. He went to investigate the sound and what

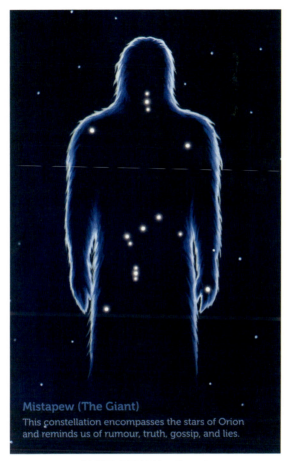

Mistapew (The Giant)
This constellation encompasses the stars of Orion
and reminds us of rumour, truth, gossip, and lies.

MISTAPEW (the Giant)

he discovered made his hair stand on end. On the edge of the
creek he was near… by a huge birch tree… laid a strange looking
humanoid form! Mistapew… Sasquatch!

It was a young Sasquatch… being small and totally ripped
up. It would seem it had injured itself somehow and the wolves
or another animal had tried to eat it. Somehow it would seem
the young Sasquatch had fought off its attackers but was mortally
wounded. It lay there bleeding and moaning.

My grandfather comforted it as best he could but there was nothing he could do. The Sasquatch passed away in his arms.

He buried it beside the large birch tree and said a prayer.

When he returned home he told my grandmother and they agreed not to mention it to anyone.

Somehow the story got out and my grandfather was paid a visit by some business people from the town of The Pas. Money was offered for information as to the whereabouts of the large tree, portage, and creek.

My grandfather took this information to his grave. No one ever knew the location of the young sasquatch.

It was said that after this time, up until the hydro development projects flooded him out, his traps were always full. Grandmother said this was because of the kindness my grandfather had shown the young sasquatch. Mistapew.

Another name given to this group of stars (Orion) that is associated with the shape shifter, *Wesakaychak*, is the Winter Keeper... *Pipoon Niki Titi Cikiw.* This constellation arrives in the night sky as it turns cold in our northern hemisphere. Pipoon Niki Titi Cikiw stays with us all winter long and as our northern hemisphere warms up, Pipoon Niki Titi Cikiw catches up to the sun and is gone from our vision until it again turns cold in our reality.

We are told Pipoon Niki Titi Cikiw watches over all the spirits during this time, identifying the sick, weak, and troubled. Some the Winter keeps... some are taken. We do not know how this is determined, only know we should prepare for the long cold winter storms.

Another story connected to this incident concerned my father, Pinesis (Birdy). My father was a full-fledged alcoholic, as were many displaced men of this era. He drank until he passed out, woke up and drank some more.

One particular winter's night he was drinking at one of the local bars and at closing time he was thrown out, quite inebriated. He started staggering home to Umpherville Settlement, across the

Wesakayckak (The Trickster/Teacher)
This constellation encompasses the stars of Orion
and tells the legend of Wesakaychak.

WESAKAYCHAK (the Everywhere Long Ago Spirit)

frozen Saskatchewan River in a fierce blizzard. At some point, staggering across the river, he fell and could not get up.

The last thing he remembered before he passed out was two large arms reaching out of the swirling, blowing snow lifting him up. Whoever it was, was wearing a large, heavy fur coat.

The next morning he awoke laying on the floor in his house with a fire blazing away in the wood stove. He does not know how he got there. Eventually my father drank himself to death.

We are told Mistapew is one of the shapes the master Shape Shifter Wesakaychak slips into when visiting the physical world. Wesakaychak was given the task of naming everything in Ininew Aski (Cree World). With this job came the perk of being able to change into everything that was named. Thus Wesakaychak, being mischievous and always hungry, played all kinds of tricks and got into all kinds of predicaments.

On another occasion I was told Mistapew saved my son Colin. He said one time when we lived in Opaskwayak he got drunk and stoned and was trying to walk home from a party in a raging blizzard. In the blowing wind he got disorientated and did not know which way he was going and eventually fell down in a snow bank. He said he remembers passing out but the last thing he saw was a pair of huge arms lifting him. He said whoever it was must have had a bear or buffalo robe on because all he saw was hair. The next thing he knew, he came to his senses and he was at the doorstep of his friend's place!

MISTAPEW/MIKAN OCISIYUWIN (Fools the Wolves... Castor and Pollux)

The old man then made a wooden stake and looked around, found what he was looking for and then drove the stake into the ground. He then tied a piece of string to this stake and paced off about seven to eight paces from the centre. Here he stuck another stake in the ground. Being the little snot I was I said, "What... no tobacco?" He looked at me and said, "You just watch and listen."

I drove stakes, at equal intervals, into the ground in a wide circle around the central stake the old man had first placed. Then I was instructed to trim the willows but leave the upper portion of the leaves and branches intact. I was also instructed to sharpen the bottom ends of the willows. We placed the sixteen willows at each hole in the ground I had made with the bar and then we inserted the willows into the hole... but first tobacco was placed into the holes and prayers were said!

The next step was to dig a hole at the centre of the circle we had just made. The pit was about two feet wide and one foot deep. I was instructed to place the earth from the hole I had begun to dig at a point of the old man's choosing and shape it into a small mound.

We then bent the willows and fastened each on to an opposite willow on the opposing side of the circle. We had created a dome of willows which we then proceeded to cover with the canvas tarp. Once this was done I was sent inside the covered dome and told to wait and let my eyes get adjusted to the lack of light. I was then instructed to see if there was any light showing anywhere in the dome. If there was light I was to let him know where it was and he would adjust the tarps and block it out. When it was totally dark I was told to crawl out.

The next order of business was to go and haul wood for the fire. We drove to the gathering's large wood pile and the old man chose lengths of logs which I loaded onto the truck. I was then instructed to take these back to our site and cut them to length and split some of them.

I said, "Why don't I just take cut to length, split wood from the pile already here and we can use these for the fire?" The old man looked at me and said, "and what would you learn from that?" I did not have a response to that so I did what I was told… grumbling all the way.

So, I spent all day working for this old man and came to realize we were making a sweat lodge! I had heard about this ceremony and participated in one before but never knew how much work was involved. I had just shown up and gone into the ceremony. I began to appreciate the work involved and found that all this effort got me out of MYSELF and all the sweating with hard labour began to clear the fog of alcohol and drugs I had been living in. I began to look forward to the actual ceremony, thinkin', "this is going to be good!"

The rocks were laid down and again tobacco was placed and prayers were said! The rocks were covered with wood and all was just about ready for the ceremony. I then went to get water for the ceremony and as I was doing this I stopped by my gear and picked up a towel and some shorts. I was getting excited and looking forward to the lodge. When I got back to the site I was instructed to light the fire and make sure the rocks were covered at all times and never was I to leave the fire untended. I was told to put tobacco into the fire and pray… talk to the fire I was told! This old man was nutty as a fruit cake!!

I was told the fire needed to heat the rocks for about two hours before they would be good to use for the ceremony. By then they should be red hot. Never leave the fire unattended and make sure the rocks are covered at all times. When people begin to arrive tell them they need tobacco if there is anything they are wanting to ask about inside the sweat lodge. These were my instructions.

The old man said he had to go and get his stuff for the sweat and off he went. About an hour and a half later people began to arrive for the lodge asking all kinds of questions that I had no answers for other than tobacco and thirty-two rocks. By this time the old man returned carrying a suitcase and his sweat lodge wear… towel, shorts. He asked me to get a smudge of sweetgrass going and smudge the lodge, all the people, and then give the sweetgrass to him. He began to set up his pipe and medicines and placed them on the altar.

Finally everything was ready for the ceremony. The old man told me to tell everyone that the lodge would start shortly and that everyone should get themselves prepared. I rushed to change into my sweat shorts and grab my towel. The old man looked at me and said, "Hey what you doing?" I said I was getting ready for the ceremony. He said, "You are to watch the fire, open and close the door when I tell you, and bring in the rocks when I tell you, and otherwise DO what I tell you… you are the door and fire man."

This was the very first time I had built a sweat lodge and pre-pared for the ceremony and I appreciated the work involved. I never went in but all the work focused me and got me out of myself. I felt so much better when I finally hit the pillow that night.

Usually people who are attending a sweat lodge just show up at the appointed time and participate in the ceremony. Very rarely is the work and preparation of how the lodge, rocks, wood, tarps, etc. etc. get to be there thought about. This fact is really evident when organizations ask for lodges from an Elder and are surprised when they are made aware that $100 does not cover it. People need to be cognizant of the fact that there is a lot of work and cost involved. This does not include the five to six hours that two to four people take from their day in order to get the ceremony ready.

I still went back to the streets, to the drugs, alcohol, and all that went with this lifestyle and it was not until a good (or bad) ten years later that I finally realized what that old man had shown and taught me. The old man's name was Ernest Tootoosis from Poundmaker Reserve in Saskatchewan. I never got to talk to him and thank him for this knowledge he shared and I heard he had passed on by the time I decided to leave that self-destructive lifestyle. Prayers for my teacher.

Early in the winter of 1974, I arrived back in The Pas, Manitoba from one of my many journeys through the "skid rows" of Western Canada and found things had changed on the "rez." Some youth from Opaskwayak journeyed to Ottawa for a rally/protest with "The Native Peoples' Caravan" and brought back with them ideas about our culture. These youth were people who I grew up with for the most part and they wanted to start a pow wow group on the rez.

This is where I began to learn about other aspects of our culture and began pow wow singing. I loved this so much I began to realize I wanted to do this even though there was no drugs or alcohol involved! This pow wow group took us all over Western Canada singing and dancing. We were invited to schools and communities throughout northern and central Manitoba and we became pretty good at it. After a couple years of rough going we began to be acknowledged for our commitment.

Thing was, we were acknowledged by others but not by our own community. Our people saw us as a novelty, ridiculed us, and the youth would ask us if we were "real Indians." We took this in stride and continued on. We travelled all over Western Canada winning singing trophies and being asked to be host drum at various pow wows. I remember one such time we were asked by actor Gordon Tootoosis to be host drum at a large pow wow he was putting together in Ottawa. We drove from The Pas to Ottawa, Ontario in about thirty-two hours. There were three carloads of us and we had a whirlwind trip. We were stopped by the Ontario Provincial Police (OPP) twice for looking suspicious.

One of our members was Isaac Bignell, a well-known artist, but at the time he was just finding his talent. We called him "Shingwak" because he reminded us of a character from an old CBC television show called *The Forest Rangers*. He was always laughing at something and saw humor everywhere. I was sorry to hear he had left us for his final journey.

We had some good times as we travelled from here to there and back again. All this made me feel good about myself, but not good enough to leave the self-destructive lifestyle behind. Always went back and things got worse and worse. I finally left the pow wow group because it was getting in the way of my lifestyle.

As time went on I began to see I was surrounded by ass-holes! They were always getting me in trouble. Case in point: I was drinking in a bar in The Pas and got into a big fight with my girlfriend at the time. We made a big scene in the bar and I left, swearing and cursing at all those ass-holes! I bought some beer at the vendor and intended to go to the rez and party. As I walked out of the vendor with my beer I rounded a corner and was met by a fist. Down I went. I was out cold! I came to, laying on the sidewalk as a snow storm blew up. Fucking ass-holes I thought… I got up, dusted myself off, checked my pockets and found no money! Ass-holes strike again!! I decide to walk back to the rez in the growing blizzard. As I arrive at the ice-road, which is the shortest way to the rez in the wintertime, I start walking across… feeling sorry for myself for being surrounded by ass-holes. Next thing I know the ice gives way and down I go into the water! I am screaming, thrashing and clawing at the edge of the ice. Finally I get a grip and crawl up out of the water onto solid ice! As I rise to stand up the ice gives way again! Down I go again! Again I fight for purchase, screaming, crying, swearing and again I make it out of the water. Fucking ass-holes!! This time I crawl away from the hole I had made before I try and stand up. I am alive! I am standing in a blizzard as my clothes freeze solid and I know I have to make it across the river to the nearest house. Somehow I make it and am helped inside. A party is raging! I make myself at home and all is forgotten. The only thing I remember is my ass-hole girlfriend and her ass-hole friends. What a bunch of you-know-whats.

Again Death reminded me of the road I was on. Still… never sunk in… and the beat goes on… tada-dump… tada-dump.

It seemed I could not shake these ass-holes. Everywhere I went they showed up. Everything I did they were there. One day it slowly started to dawn on me that maybe, just maybe these ass-holes were NOT ass-holes. Maybe the ass-hole I saw everywhere was me! Could this be?? Could I be the problem? NAW! Could not be ME!! I was the victim, the poor picked on little Indian boy.

Near the end of my drinking, drug-filled life, I would hallucinate at the drop of a hat. Get a little drunk, smoke some grass and I was off. I would see big crowds and be involved in large parties but none existed… only in my mind. I would see wild animals, people would change before my eyes. I thought… "maybe I need some help." Nahh… I'm OK.

The summer of 1981 found me working in Winnipeg, Manitoba. There were three of us from back home working for a criminal justice organization. Every payday we would book days off and head back to The Pas. This one particular time we rented a car and decided to drive back home and party. We left Winnipeg with grass and whiskey on a fine sunny morning. By the time we got to a major junction two and a half hours out of The Pas, we ran out of whiskey and decided to take a 30 kilometre pit stop and reload on whiskey and beer. We ended up in Grand Rapids at the local bar. One thing led to another and we got to drinking and partying with the local ladies. I left my buddies to go to a party and next thing I know the people at the party are telling me I had to leave right away. Some locals heard we were in town and that we were from The Pas. We would not leave Grand Rapids unscathed. I guess the locals got wind I was at a party and were going to "fix me up good." They came in the front door and I left by the back window.

I began hitch-hiking down number 6 highway north. I figured I would take the long way home. As luck would have it I got

a ride by a guy whose sister I knew. He asked me where I was off
to and I told him anywhere but here. We got to talking and he
said he was working at a mining camp and was selling acid on
the side. He asked me if I was interested in buying some acid. I
said, "How much?" He looked at me and asked, "How much you
got?" I thought yeah right… as if I am going to tell you how much
money I have. I dug in my pocket and pulled out about twenty
five dollars in bills and change and said this is all I got. He looked
at me… dug in his pocket and pulled out a handful of colored
paper he said was blotter acid and gave it to me. I thought yeah
right you are going to give me all that acid for twenty five bucks.
I gave him the money, looked at him and downed the acid, all in
one shot. He looked at me and said, "YOU ARE CRAZY!" He
let me off at a gas station and continued on his way.

I waited for a bus deciding I was going to go back to Winnipeg.
I caught the bus and as we were riding along the acid started doing
its work. I began getting very paranoid and very, very claustro-
phobic. I went running up the isle of the bus and started yelling at
the bus driver that he had to stop and let me off. RIGHT AWAY!
The bus driver I'm sure wanted me off the bus just as bad as I
wanted to get off the bus so he stopped.

So there I was standing on the side of the road in the middle
of nowhere. It was dark and I was standing in the centre of a world
spanning boreal forest hallucinating on acid! As I stood there I
could feel the wind and hear the roar of the big trucks as they went
flying by. I was freaking! At some point I must have run into the
bush. It was at this point I lost my mind.

I find myself in the forest, laying on my back looking up at
the night sky. I begin to see all kinds of things in the sky. Bears,
wolves, moose, birds and a horse were running around and flying
up there. In an instant I see my body lying in the moss on the forest
floor as my spirit floats up. I rose above tree level and saw the vast
night sky. I looked down at my body lying there in the moss and
watched it dissolve into the earth and then my spirit flew.

What happened next is still vivid to my memory. My spirit flew back through time and space and I re-experienced everything I'd ever done. All the things that had happened to me to that point, I relived. All the crappy things I had done to people, all the hurt and misery I had caused, all the pain and suffering I had instigated and all the times I had used peoples' feelings and emotions just to get what I wanted… I revisited over and over and over. During all of this, there was a pain running through my body that my mind said I could not do anything about and had to endure.

As this went on for what seemed like forever, I began to realize that this is what it means and feels like to be condemned. My spirit would relive all the trauma I had experienced and had caused. I was stuck in a recurring loop of time and space and I was given insight as to what I could have done better or not done at all. I slowly began to realize I WAS DEAD!

I would relive this life experience forever and ever and end up at this place… in this way. I remember screaming, crying… "I don't want to die… I don't want to die… I don't want to be dead!" Again this went on forever. Then at some point I realize I am in a thick, thick white fog, screaming and crying… "I do not want to be dead." As I am floating through the fog I begin to notice a blinking light through the heavy cloud. I look and continue screaming as I slowly move toward the light. Again this seems to take forever but it seems I am getting closer to the light. I blink and in that moment all the thick, heavy fog has disappeared and I am standing in a forest!

The bright summer sun is high in the sky and I hear every sound that surrounds me. I hear the bees, the flies, the numerous birds. It is so intensely bright that my eyes are tearing up and I have to squint to see anything at all. I realize I AM ALIVE! I AM ALIVE!!

As I stand in the forest taking all this in, I look at my hands and see that they are all swollen. Puffed up like marshmallows!

My arms are the same... puffed up! I feel my face and ears and my face and ears are in the same state... PUFFED! I am inflated! I then realize I am also lost... in this forest. I began to panic but remembered I had been in this situation before and calmed myself down. Don't panic... have some bannock. I never had no bannock but had experience of what to do...

I sat down in the forest and began to look at my shadow and those the trees made. As I was doing this I began to hear the distant hum of cars and trucks every so often. I listened for the sound and estimated in what direction it was coming from. I began to walk. I found my way to the highway and began to hitchhike. I looked at the sun and decided to head north. As luck would have it in a little while I was picked up by a missionary who was headed my way. He began to explain to me about the evils of drugs and alcohol. I thought, "hummm, maybe someone or something is trying to tell me something here." I never got it...

I arrived back in The Pas from my great adventure and dove right back into the drugs and alcohol... but with a story to tell. It would take me another year or so of more confusion, anxiety, worry, anger, hate, and shame before I left the streets behind.

Years later I took some tobacco and cloth to a sweat lodge and asked an Elder what had happened to me that long lonely summer night when my spirit left my body. Here is what I was told. The Elder said,

> You died. Your heart stopped due to the shock all the LSD put on your body. You wandered in the spirit world reliving your negative actions that YOU knew were wrong. The dull pain that you felt, while all this was happening, is what brought you back to reality. That pain was what saved your life. The pain was caused by *sakimayuk*, mosquitoes. The mosquitoes had a feast on your blood as you lay on the forest floor. All the blood that was taken out of you extracted the LSD and in its place put

mosquito venom into your body and bloodstream. Thousands
of mosquitoes feasted on you and put their medicine into you.
They did two things that saved your life. First, they extracted
the LSD from your bloodstream and body. Next they put their
venom into your bloodstream and body. This saved your body
and mind. First they took the poison from your system. Then
they put venom into your system. The vast amount of venom
that went into your body and bloodstream must have shocked
your heart to restart. You were sent back! I guess there is something you still have to do!

Let me tell you, there must have been thousands of stoned
mosquitoes in the bush that night! So that is why you came out
of the forest all PUFFED UP…mosquitoes. That is also why you
came out of the forest at all."

I was saved by mosquitoes! Little bastards…

So the answer to the age old philosophical question "If a Cree
falls in the forest… does anybody hear?" the answer is no. That
is if the Cree does not survive, but if the Cree makes it out of the
forest, then anyone within hearing distance will know about it!

During this wandering time I had been employed at numerous
jobs. I remember one aspiration I had… this one job I really wanted…

When I was younger, filled with anger, shit and vinegar, I pursued
a "dream job" that I thought would be so awesome to have. That
job was working on the Canadian National Railways Extra Gang
as a labourer. My great ambition was to work all summer and get
enough weeks in so I could draw unemployment insurance all
"freeze-up," winter and spring just to get my job back in the early
summer and do it all over again! The CN Extra Gang component
of the CNR was established to repair and maintain the railway
tracks and infrastructure needed for the smooth operation of trains
and cargo on the move.

Those of us that worked at this profession lived hard, drank lots, fought at the drop of a hat, forgave with a bottle of beer, cheated, stole, and lied to each other constantly. We were brothers!

We drank every chance we got. We would spend all our earnings in one weekend of debauchery only to wake up Monday morning in our railroad bunkhouses, broke, beaten, battered and sore. Then we would work like crazy for two weeks and then do it all over again! In times of clarity I thought we were in rough shape but, there is always someone in worse shape somewhere. This one time we were partying in a little village called Waboden, Manitoba where our railway bunkhouses were parked for the time being. Everything shut down in the small village way too early for our liking so we decided to go to Thompson to continue our party. Four of us hired a taxi for the 106 km trip to Thompson. We were driving down the highway feeling pretty good and looking forward to Thompson GIRLS, Thompson DRUGS and the infamous THOMPSON INN.

As we drove along we noticed a semi-trailer parked on the side of the road with the driver standing in the middle of the road waving. We pulled up and stopped to ask what was wrong. The driver looked at us in a daze and said, "Is this the Thompson Highway?" We looked at each other and burst out laughing. Is this guy for real we said? We told him yes and he turned to go back to his truck… then he stopped. He slowly turned around, came back to the taxi and asked us, "Which WAY is Thompson?" We totally killed ourselves laughing! We told him he better get some rest right where he was and worry about things in the morning! We left him there but did we have a story to tell. We thought, What a L-O-S-E-R! then continued on with our debauchery.

One of the people I worked with on the extra gang I would run into in another lifetime, in another setting, in another frame of mind, in another reality. At that time he was known to us as "Neckbones."

As time progressed I drank myself out of my dream job…

When I start remembering some of the situations I found myself in I am convinced that without the strong warrior spirit that resides in some of our people, a lot of us would not be here. This spirit is evident in people like Dennis Banks, Leonard Peltier, Harold Cardinal, David Courchene Senior, my brother Larry Morrissette, to name a few.

There are many other of these warrior spirits that accompanied the everyday person such as yourself or myself in our everyday existence. One such person was/is my friend Patrick Shingoose. Pat saved our hides a number of times. It was at these times when the majority of us would be resigned to the fact that the inevitable is going to happen and get ourselves ready for whatever it was that was going to happen… to happen. Not Pat. There are two occasions I can remember clearly when this was the case.

During the 1970s in the town of The Pas, Manitoba, a lot of racist animosity was happening. Our people were being attacked on the streets of The Pas and of course retaliations would spring up.

One evening Pat and I were walking to a local bar, coming from a party in the southeast part of town we called "the ghetto," one of countless housing projects throughout this land where our people are herded into.

As we walked down the dark street, two cars full of whites pulled up yelling and screaming, "there is going to be some scalping tonight!" I thought, "Oh shit! We are in for a beating now!"

They jump out of their cars with baseball bats and two by fours. They were ready for this planned attack on anyone who was INDIAN.

I have already resigned myself to the fact that we are going to be beaten black and blue… or worse. There was nowhere to run. We were surrounded against a wooden fence by numerous hoodlums with bad intent.

So we are standing on the sidewalk backed up against a fence, surrounded by young whites all drunked up and I am waiting for the Charge. Pat nudges me and rips a wooden picket off of the fence we are standing against. I do the same and next instant WE are the aggressors NOT the victims. We chase them around the street until they jump into their cars and skedaddle!

I had a story to tell!

Another instance happened on the streets of Winnipeg in the early '80s. Again, Pat and I were returning from a party, on our way to the next party, in the early morning hours. We had a 12 pack of beer… walking down a dimly lit street when all of a sudden two carloads of young white guys pull up. One had a baseball bat! Again they were yelling and screaming, "time to hunt some dirty redskins" or something to that effect. Anyway, once again we find ourselves surrounded but this time there were no picket fences to save us! The enemy advanced with bat held high. I was backing up slowly and I heard glass break. I looked around and there was Pat with two jagged broken beer bottles, one in each hand! He had taken the beer out of the case and was now using these as weapons!

I heard screams from our adversaries, "Holy Fuck, he has broken bottles!"

So immediately I grab some bottles and break them and start waving them around. Low and behold, once again we are the attackers NOT the attacked! Once again, the white guys are sent running with their tails between their legs!

Again I had a story to tell!

My brothers and sisters reading this can surely identify and name someone who has been touched by "the warrior spirit" and saved our asses on various occasions.

In October of 1982 I left The Pas for the greener fields of Vancouver. The walls were closing in and everyone could predict my lies and had heard all my sob stories. Time to leave… Ramble On…

I arrived in Winnipeg with a plan. The first part of the plan was to PARTY. Then when that started to go bad I would continue on to Vancouver. I drank and did drugs from late October, 1982 until mid-January, 1983. For a large portion of that time I was in a total mind-numbing, spirit killing, don't give a fuck, messed up black-out.

The only portion I vaguely remember is shopping at Safeway. It must have been before Christmas because I had turkeys, ham, steaks, stuffing mix, salad stuff… everything one would need for a Christmas feast. I filled up the shopping cart and just walked out of the store! I walked down back lanes to my auntie's place. I dumped all the food there and then continued on to a party. Lost all memory of anything else. I was wandering in a thick dark fog….

I finally came to consciousness, sitting in a barroom at the Mall Hotel, in Winnipeg, on a cold January day, holding a glass of beer, feeling totally sick. Sick of drinking, sick of doing drugs, sick of stealing, sick of lying but most of all… sick of myself. Time to go to Vancouver.

First I needed to sober up and get some money that did not involve stealing or selling drugs. I had heard that if a person was trying to help themselves get straight there were certain agencies that would help. So my plan was to sign in at the local Alcohol Foundation of Manitoba (AFM) to do a treatment program. If I did this then I could apply for welfare and get some cash so I could get to Vancouver. All I had to do was tough it out for 28 days. No problem…

|||

On January 23rd, 1983, I went into treatment for the 20th time. But this time… something happened that changed my life. I met an Elder by the name of Bernelda Wheeler, a Cree woman. She told me something that shifted my universe. After what she shared, I saw the world in a totally different way.

Now I had entered treatment numerous times before, always with an ulterior motive. I endured all the self-help exercises and endless catch phrases but nothing sank in other than the fact that I was fucked-up… but I already knew that.

Bernelda did not preach, she shared. She did not attack with rhetoric, she spoke from the heart. She did not tell you what you should do, she made it clear the choice was yours but this is what *she* did. She did not mention Jesus or Christianity, but a higher power. For these reasons my ears finally heard something that made sense to me. She talked about the "medicine wheel" and how we were medicine wheels.

This was not the first time I heard the term "medicine wheel" used. But this was the first time I actually related the teachings to my experience. The first time I heard of the medicine wheel was at a youth/Elder workshop and a man named Hyemeyohsts Storm was talking about his new book called *The Seven Arrows*. I listened because this was some wisdom originating from our people. At the time I did not apply any of the teachings to my situation but thought they were cool.

Bernelda spoke about how we carried our own medicine wheel with us where ever we go. We sit at the centre of that medicine wheel and interact and influence and in turn get influenced. She said, "me, as an alcoholic carried this wheel around with me and unconsciously sought ways to re-enforce this image I had of myself. I thought I was bad, useless, a thief, liar, drug addict and alcoholic therefore these are the people I sought out wherever I went. I felt crappy about myself so I was attracted to like-minded people. We helped re-enforce our situation. We all felt crappy about ourselves so we did things that were crappy. We lied, we stole, we hurt before WE were hurt, we hated, we used people and we lost ourselves in alcohol and drugs."

She went on to say, "how we feel about ourselves reflects on who we associate with, what we do and how we live or where we go…. These four points all feed off each other and capture us until at some point we get sick and tired and cannot do it anymore… so we change… or we die."

"We All Carry a Medicine Wheel With Us…Wherever We Go."

Where You Go

Who You Associate With

YOU

How You Feel About Yourself

The Things You Do

The wheel we make is applicable in whatever we do… be it negative or positive.

What she said made so much sense. She had opened my eyes, unplugged my ears, engaged my mind, and melted my ice-cold heart. I was human again!

She said, "In order to walk away from the lifestyle you led up to this point you have to change your entire world. Since everything is dependent and feeds off of each other on your medicine wheel… you have to build yourself a new wheel. You cannot just change one or two things… you have to change everything."

It seemed like a bright light was turned on in my dark universe. Hope was everywhere. I walked around for the rest of my stay in treatment with a big dumb smile on my face. It seemed that everything had a golden hue or aura to it. Thus I began my walk into sobriety.

When I left treatment I was told to attend Alcoholics Anonymous (AA) meetings. I was also told to keep myself occupied because the time I had spent with alcohol and drugs previously, now had to be filled doing something else. I decided to go back to school. Up to this point I had a grade seven education. I had tried various courses throughout my drinking/drugging career but it was a haphazard attempt and I never really finished anything I started. I had started grades eight and nine at one point but found I was too cool for school. Now I found I was unemployable because I was uneducated. Well I guess I should say I was employable but only for basic manual labor.

Before I decided to go back to school I began working at Casual Help All. I was sent out early every morning to do all kinds of simple, dirty, hard manual labor tasks for barely minimum wage. It all came to a point one cold morning when I was sent to Assiniboia Downs, the local horse race track. I had to clean out horse stalls. I was shoveling horse shit and washing down horse piss. As I did this I asked myself, "Do I want to be doing this for the rest of my life?" The answer was no!

I walked away from that job and enrolled in grade nine classes at the Winnipeg Adult Education Centre on Vaughn Street in Winnipeg. I spent the next sixteen months finishing my grades nine, ten, eleven, and twelve. I was finishing grade twelve courses in June of 1985 when someone suggested I apply at the ACCESS program at the University of Manitoba; I said sure why not.

I went to a place called the Winnipeg Education Centre, where the off campus ACCESS program was being delivered, to apply for university. I wanted to get a degree in social work as a way to work with my people. As I walked into the building the place was bustling with activity. I asked someone where they were taking applications for the Bachelor of Social Work program. When I arrived at the room, it was crowded with about sixty people. I asked someone how many spaces are being offered and was told sixteen. I thought, "well so much for that wild dream." I walked out of the room intending to go home and as I walked by another room I noticed about twelve people lining up at a table. Being curious, I walked in and asked what was going on. I was told people were applying for a Bachelor of Education program. I asked how many spots available and was told sixteen.

I thought to myself, "so much for social work!" I think I am supposed to be a teacher!

SIXTEEN! I could do this! So in a sacred manner I found my path to education. I was destined to become a teacher! So this is how I got into the education program and got a Bachelor of Education degree (B.Ed.) from the University of Manitoba.

During the time I left the treatment program until I found my own place, I stayed with my aunt Jean and my uncle Frank. It was hard going during this time due to the changing of my lifestyle and the fact that my aunt and uncle were always drinking. I was told if I wanted to stay sober I needed to go to AA meetings and this I did… religiously. A meeting every evening and three times a day on weekends. I became involved in AA. It became my life. Initially I helped wherever I could, setting up chairs, making coffee, getting the meeting room ready for meetings. I got involved in the administration of the AA chapter I was going to. I helped with other people trying to recover from their addictions. I chaired meetings and eventually began being invited to speak at meetings, first at our group and then at other groups around the city. For the first two years of sobriety I was an AA fanatic!

On top of the meetings I began to socialize with other AA members outside the meeting rooms. We formed a fastball team and played baseball full tilt. Sometimes we would play four to five games on a weekend and two or three games on weekdays in the spring and summer evenings. I started to get back into shape. I also began jogging at this time.

This kept me focused and busy because when someone leaves a life, a lifestyle, a way of thinking, doing, reacting, feeling and adjusting for something completely new... it is scary. There is a hole where alcohol and drugs used to be. A hole that companionship, comradery, sex, numbness, and anger used to fill. Now I was hanging on the edge of darkness... looking for the light. This is when I was saved by Pawamiuk... dreams...

I began to have vivid, lucid dreams. I did not know it at the time but this was a healing, inspirational, motivational, and spiritual process that was given to me as a way to come to terms with who I was, where I came from, where I was going, and how I would get there.

My biological family was ripped apart when I was about four years old. I only knew I had an older brother at the time. It was not until much later on in life that I found out about other siblings because no one ever told me about my family. I remember one January I went to see an Elder because of some of the dreams I was having. I was instructed to Fast.

I was to lock myself away, in my little basement suite, close all the lights, shut off all communication devices and think, pray, and dream. When I began it was a stormy February day. I sat there in my suite and listened to the wind moan. I went without food and water for four days and during that time I found discipline, belief, hope, and guidance. I dreamed of spirits and ceremonial songs came to me. I remember one dream in particular. At the time it did not make sense to me, but nonetheless was very clear and remained with me for a long while. When I found out what it meant I was floored and filled with peace.

I dreamed of four spirits. They were bright yellow, green, blue, and red. They shimmered as they danced around me. They made me laugh, cry, and feel at peace. The yellow spirit flew right up to me face to face… looked me in the eye and said, "Elda Rose" and I awoke with a start. It was so real… High Def…even! A spirit calling song was given to me at that time. This was the very first Fast I ever did.

After the Fast I continued moving forward with my sobriety, education, personal and spiritual growth. A couple of months later I was visiting my aunt and uncle, Jean and Frank, and mentioned about the Fast I had done. I spoke about the dream and the spirit that said Elda Rose. My aunt looked at me, took my hand and told me this. She said, "You and your *nistes* (older brother) had a baby sister. Your mother was by herself raising you and your brother all the time. Your father was always gone. Your baby sister got very sick with the influenza and she did not get better. She passed away in my arms. Her name was Linda Rose."

When I heard this I began to cry and then laugh. My sister came and seen me! The sister I did not know I had! She came and gave me peace and hope! Again I felt I was not alone and what I was doing meant something!

During the second year of my new existence I was invited to a sweat lodge. I had not been to a sweat lodge for about three years to that point and was eager to reconnect. I was asked if I knew how to build a sweat lodge and I remembered all Ernest Tootoosis had taught me. Once again I dove right into this opportunity and have never turned away since.

So now I had school, AA, baseball, and sweat lodges in my life and these kept me very busy.

Remnants of past deeds still haunted me, but the AA step program helped bring these into perspective. I would have nightmares of things I did and would dream of being on a drug and alcohol rampage and wake up sweating and at times crying. One nightmare/dream in particular plagued me. It had to do with my father's parents… my paternal grandparents. I had been lost inside myself as well inside the monster of drugs and alcohol, and I did not feel anything for anyone or anything. When my grandfather passed away I did not attend his funeral. I was too busy chasing drugs and alcohol. My grandfather was always kind to me and would stop whatever he was doing when I would visit and talk to me. I felt nothing when he died. When my grandmother died I attended the funeral, half drunk and immediately after the funeral went and lost my mind for days. So, when I sobered up these two would come to me in my dreams and look at me. I would always be afraid of them. I would scream, cry, and yell at them that they could not do this. They were dead and gone now. But apparently they were not gone. They still lived in my guilt, shame, and loneliness. I always woke up sweating and screaming.

I spoke with someone about this and was told I should go see an Elder and ask what I could do. I did this and was told I needed to go speak with my grandparents. The Elder said, "Go to where they were buried, put tobacco down and talk to them. Tell them you are sorry and that you are trying to change your life. After you have done this, they will come and speak to you and you will no longer be afraid of them."

So I went back to The Pas, found their graves, put tobacco down and prayed and spoke with them. I cried and cried and felt so much better after this was done. The next time I dreamed of them they were smiling and I was happy to see them. I was no longer afraid. In my mind peace had been made. They come to me with dreams of hope and comfort now whenever I have dreams of them.

I found dreams came to me that offered guidance, hope, glimpses of possibilities, direction, and solace. When I moved into a place of my own I began to dream. I would have very vivid dreams. One of the very first vivid dreams I had is still very clear in my mind. It revealed itself to me like this.

During the first stage of changing my lifestyle I was very lonely and there were a number of times when I thought about going back to the street. Every week I would phone home and check in with my nistes (older brother) Peter or he would call me. The first thing he would ask me is, "Are you still sober?" He had seen me quit drinking and drugs so many times already and was expecting me to "fall off the wagon." I would tell him I was still sober.

This one particular night (after a very, hard, lonely and soul-searching day) I fell asleep and dreamed.

In my dream I went flying back to Umpherville Settlement (we called it the U.S.) where my grandfather had built a two room log house for my granny Carrie. It was twilight and the sun cast off a golden hue over the world. I landed on the front stoop just as it was getting dark out. My grandfather was standing there waiting for me. He took me inside the house and I saw my granny sitting on a chair sewing a quilt. The two of them were the only ones I saw in the house.

My grandfather then led me into the other room that was used as the kitchen. As I walked into the room I stopped in my tracks. It was pitch black in the room. I wondered why my grandfather wanted me to go into this room. As I stood there, my grandfather lit a kerosene lamp and held it above my head and slowly we walked into the centre of the room. As I stood there looking into the gloom I began to make out shapes that were shuffling around in the dark. As my eyes became adjusted to the low light I saw people slowly moving in a wide circle, clockwise. They were shuffling slowly in the dark, walking with their eyes closed and their hands and arms stretched out in front of them. As I watched this spectacle I began to see who these people were. They were all my relatives!

I saw my brother, aunts, uncles, cousins, and parents slowly walking in the dark with their arms outstretched. After a time of watching my relatives shuffle in the dark my grandfather said, "*How... ekwa nosisim*" (time to go grandchild) and I knew we had to leave. As I left the darkness of that room I wildly reached out and grabbed someone. I did not know who it was but I held on to them and dragged them out of that room with me. Next thing I am flying away from the log house. I look back as I fly away and see my grandparents standing on the stoop waving at me. Standing beside them is the person I had grabbed as we left the dark room. It was my cousin Joe Pelly. The three of them were standing there waving at me as I flew away. I open my eyes with this dream still very clear in my memory. I feel calm and hopeful. I go about my business and forget about the dream.

The end of the week (Sunday) finds me calling my nistes (older brother) and letting him know how I am doing. As we end our conversation my brother says, "Hey I got some news, our cousin Joe Pelly has stopped drinking and doing drugs. Yeah, he is now a born again Christian. He has found God!"

At this news, the dream I had comes rushing back and I see my grandparents laughing at me, all happy! I begin to cry and thank my brother as we say goodbye. I feel blessed... touched by the sacred. I have hope, peace, direction, and commitment on my decision to walk this road. I have help. I am not alone!!

For the longest time I did not cry. I numbed my spirit, tortured my soul, starved my emotions, and froze my heart. I was the cold winter wind that took, took, took and never gave. I left hurt and destruction in my wake.

I lived in a world that demanded I be immune to feeling. No

connections, no loyalties, no roots, no emotion. I showed a don't-give-a-fuck face to the world.

I found in order to change my world I needed to change this aspect of myself. This was one of the hard parts.

When sobriety found me I began to thaw and my spirit cried for warmth and my emotions cried for food. It was during this time I was invited to a series of Youth/Elder Workshops that happened to be in the area. I kept coming across this gentle little man who I was awed by.

I also found it very odd that this small quiet man had a son that was loud and larger than life. I'd met the Elder's son first many years prior. When we would be on the pow wow trail we ran into a lot of interesting people. This one man we came across out west. I think it was in the Qu'Appelle Valley where we first met Walter Bonaise. He heard us singing and asked if he could sit in with us. He was loud and always teasing someone. He is a very good singer. Travelling around, whenever we ran into Walter he would come and sing with us. He was funny. So I am attending this Youth/Elder gathering and who do I see assisting an Elder but Walter Bonaise. I go over and say hello and he introduces me to his father Alex. Walter is acting very subdued. Not until later when he is away from his father do I see the Walter I know make an appearance. Very strange.

As I keep running into Elder Alex Bonaise I am in awe of the sphere of calm he seems to surround himself with. I sit with him and feel at peace. As I watch this man conduct his business I see him cry when he talks about the sweat lodge, the pipe, the earth and the people. He shows rather than tells me it is OK to cry. If it is something you believe with all your being then let the emotion, spirit, energy, and tears flow. It is a healing process.

For about two years after the first tears of spiritual emotion flowed from my eyes; I cried at the drop of a hat. I cried everywhere. I cried at movies, pipe ceremonies, pow wows, AA meetings, feasts, and sweat lodges… don't get me started with sweat lodges!

I thank this quiet, gentle, little man for showing me the tool of crying and the gift of healing.

Since this time I have been shown and made to understand that there are natural ways of healing given to us that the body uses to try and right itself. When we sneeze, this is our body telling us it is trying to adjust to something in the air that is getting into our body's operating systems. Coughing, shivering, throwing up, sleeping, laughing, crying, sweating, runny noses... these are all ways of our body trying to heal itself. Even farting and diarrhea are healing processes. There is an amusing story of Wesakaychak and the shutting down of his rear end which reminds us of how important seemingly little, taken-for-granted things can affect us (I will leave this up to the reader to look into the world of Wesakaychak and all his antics).

There were a number of different Elders and teachers placed in front of me as I made and make my journey through life. I am sure there are many more waiting along the path for my arrival. Some are old, some are young, some are smart, some are dumb... but all have something to give, show, teach, or tell. Even something like "America's Funniest Home Videos" is a teaching tool. Examples and teachings are all around us, we just have to take the time to look and recognize them as such.

I remember this one time I was having a particularly hard time dealing with and accepting what some people in the community were saying about me and how I conducted my affairs. Word got back to me that so and so was saying something about this, that, or the other thing on how I talked with people, the advice I gave, how I carried myself, and the way I did ceremony... whatever. I lost sleep over this. I *worried* about this. I *stressed* about it.

I happened to be helping at an Elders' gathering during this time and one morning as I came into the gathering, I decided to speak with an Elder about what I could do about this. I brought some tobacco with me and that morning I decided I would ask to speak with the first Elder I came across about my situation. Off I went with tobacco in hand.

I walked into the school where the gathering was that morning and as I rounded the corner to the meeting hall, there sitting at a table was an Elder. To this day, in my mind's eye, I can still see him sitting there. His name was Art Solomon, an Anishinabe Elder. I approached him and asked if I could speak with him about something that was troubling me. He said, "Sit down and I will listen to what you have to say." I passed the tobacco and he accepted it and I sat down and began to speak.

I told Mr. Solomon what I was worrying about and asked him what I should do about it. He sat there with tobacco in hand, listening to what I had to say. When I was finished he looked at me for the longest time and then he began to speak. He said, "One of the biggest problems we have in our communities is a little thing called gossip. Gossip can kill people. When you hear about BAD MEDICINE, the only bad medicine is gossip, but it is strong. It destroys—people, families, communities. The strongest form of gossip is the kind that is planted as a seed of doubt, criticism, and added to that, it slowly grows and for some people it becomes fact. Be aware of this but do not be eaten by it. Do not get devoured."

"You know what you do about these kind of people?" the Elder asked. I said, "no." He said with a sense of finality, "FUCK 'EM!"

"No matter what you do or say there will always be someone gonna say something about it. You cannot go about your business and expect to please everyone all the time. Do what you are instructed to do and do the best you can. Be good, listen, stay calm and be happy. If you work from that perspective you will be OK."

When I think about this advice I laugh and know this has had a great impact on my life. FUCK 'EM! A great teaching from a smart man!

In my second year of sobriety I met some other people who were looking for themselves too and we started a pow wow singing drum group. In a short while we were being asked to sing all over the city.

We even were asked to sing for the Queen of England... twice. We got to meet queen Elizabeth II and Prince Philip.

Also during this time we began attending sweat lodges and other ceremonies. One day a friend asked, "Hey lets go ask an Elder for our names." I said, "That sounds like a good idea. What do we need to do?" So I was given instructions on what to do and arrangements were made to see the Elder and ask about our spirit names. When the time arrived we set out to the Elder's place. He lived just outside the city and as we began our drive a blizzard started to blow. We ended up on a snowed-in highway stuck. No matter what we did we could not make it to Elder John Stonechild's place. When we finally got back to the city my friend called the Elder and explained what happened and new arrangements were made.

This time the Elder said he would come to us. So once again we made preparations, got everything ready, and waited. Again we did not get our names. The Elder was called in to work on some kind of emergency (he worked at Stony Mountain Penitentiary). So once again my friend began to make alternate arrangements. I told my buddy, "You know... things have not worked out a couple of times so maybe I am going to let this go and when I am ready a spirit name will find me." He continued on and got a spirit name. I decided to wait a while.

The same friend said we should go pass tobacco to his brother in law, who was a pipe-carrier and ask if he would teach us about the sweat lodge, sacred pipe, and the teachings. This we did and this is how I came to work for my mentor, teacher, counsellor, and friend... Tipiskawi Pisim Achapis Okimaw: the Moon Bow Chief... William Dumas.

We would travel to Mistawasis, Saskatchewan, a small reserve west of the city of Prince Albert, to hold ceremony with Grampa Joe Doucette and his assistant Antoine Sand. We travelled there many times during the years to sweat and fast as well as visit.

I kept myself busy with university courses, sweat lodges, AA meetings, pow wow singing, organizing coffee houses (along with a small group of AA people) and dry socials (non alcohol dances).

One day William said to me, "Don't make plans for the upcoming weekend. We have to go to Thompson to do some sweats for a gathering that is happening at mile 20." Mile 20 is ceremonial grounds located along the highway to Nelson House, Leaf Rapids, and Lynn Lake, approximately 36 km north of Thompson, Manitoba.

We left Winnipeg Friday evening and got to Thompson late. We spent the night at Sonny and Marie Ballentyne's place. Early the next morning we went to Mile 20 and began preparing for sweat lodge. During that day seven sweat lodges were done and William conducted five of them. I helped him as well as watched the fire for the two other lodges that happened that day. It was a long, busy, tiring day. But it was good work and very humbling and relaxing.

We got back to Sonny and Marie's place late that evening and made plans to drive back to Winnipeg the following morning. The drive is a good seven hours more if you stop and have lunch, washroom, and gas breaks… but we had to leave because William had to be at work on Monday morning and I had classes.

Now Sonny and Marie have children and one of them, a young daughter that took a special liking to me and I to her. Her name is Flora and she is special. She is awesome. She is intellectually disabled and thinks like a young child. At the time she would have been about eleven or twelve years old. Flora made me happy. Her joy and ease radiated out to everyone.

After such a long day I was ready for some rest. A good night's sleep was needed for the long drive ahead. I laid down and fell asleep immediately. I began to dream…

In this dream I was walking along a slow moving creek and the sun cast a soft golden glow all around me. It was twilight. As

I walked I was looking at the clear flowing water in the creek. I saw tiny white objects floating just on the surface of the water. I went to the edge of the creek to get a better view of these objects. I could not see them any better so I decided to wade into the creek to investigate further. As I walked into the stream I began to make out what these objects were. They were pencils floating vertically with the erasers acting as a float to keep them buoyant. I was waist deep in the water by now and I began gathering these pencils in both my arms. Pretty soon I had both my arms outstretched filled with pencils and I was laughing. As I was doing this… standing in the water waist deep… with an armful of pencils laughing like a madman… I heard a voice. I stopped laughing and looked around to see who it was. As I looked around I noticed a stone bridge the spanned the creek and standing on that bridge were Flora and my teacher and mentor the Moon Bow Chief, William. Flora asked me, "Who are you?" I laughed and said I am the Dream Keeper. They both replied, "Yes, yes… you are the Dream Keeper." As they said this I threw all the pencils I was holding in my arms up into the air. They flew into the sky and I saw by this time it had grown dark and all the pencil erasers began to glow. They flew into the sky and became stars and I laughed and laughed. I looked on the bridge and saw Flora and William laughing too. I woke up feeling elated with this vision clear in my memory.

At breakfast I told William and Marie about the dream I had and William said, "It sounds like Flora has found your name for you—Pawami niki titi cikiw… the Dream Keeper. This is your spirit name. This special little girl has blessed you." Marie smiled and began to speak. Marie related the story of Flora to me.

She said, "When Flora was born the doctors came and told Sonny and I that the baby had a damaged brain. They said she would be in a *vegetable* state all her life. They said it would be better for her as well as us if we just let her die. We told them NO! We took her and left there. We loved her and fed her and comforted her and taught her and watched her grow. She brings sunshine and joy to us and everyone that meets her knows she is awesome."

This is how my name found me through a special little spirit named Flora and my teacher The Moon Bow Chief. I am grateful.

This is interesting in that *Achakapis* (a contraction of the phrase achak apisis)… Little Spirit, is associated with the moon.

On the face of Tipiskawi Pisim, the moon, there is a little boy… holding two pails. My people say Achakapis is there to remind us of how we should treat others less fortunate than ourselves especially the homeless, lonely, and orphans. The head of Achakapis is the Sea of Serenity, his body is the Sea of Tranquility, the right pail is the Sea of Crisis, left pail is the Sea of Nectar and leg section, the Sea of Fertility.

I was told part of my responsibility for carrying this name was: "to hold a feast and hang prayer flags and your spirit helpers and guides (*pawakanuk*) will reveal to you your purpose." The Dream Keeper name represents not only my spirit, it identifies me as a carrier of dreams... pawamiuk and announces to ALL... past, present, and future... how I will conduct myself. All this comes down to how I feel about myself... which is what I project to the world... which in turn reflects how I treat people... which in turn reflects on who identifies with what I say, what I do, and how I follow through with my words and actions.

It all goes back to the Medicine Wheel teaching given by Elder Bernelda Wheeler. How you feel about yourself... reflects what you do... reflects who is attracted to how you feel about yourself and how you go about your business... reflects the spiritual environment you immerse yourself in... reflects how you feel about yourself...

I was also told not to run around looking for amazing, strong, awesome, in-your-face medicine people because... I would find them!

I was told to sit back, observe, see how people are treated, stay calm, feel the serenity, and ask the question, "does the talk go with the walk?" In other words use the Force... "this is not the medicine man you are looking for."

During this time of emotional, spiritual, mental, and physical growth I began writing songs and poems again as well as short stories. I submitted some here and there and some poems and short stories were published in various local papers and journals. Also was involved in organizing coffee houses at which I would sing and sometimes reveal something I had written. One song in particular got some very good feedback from a lot of people but I decided not to pursue or push the idea any further. The song was called "Where Are You Warriors?" and was written at a time of deep turmoil about the state of my family, my people, myself. I remember watching as Ellen Gabriel walked out of OKA with the little girl beside her and how the soldiers treated them. I cried...

Where Are You Warriors?

(chorus)
Where are you warriors? The people call
They are cornered by a monster, With their backs against the wall
They are calling for their leaders, To lead their pain away
They are calling for you warriors, What do you say?
They are calling for you warriors, What do you say?

Are you sitting in a barroom Playing a waiting game?
Deep inside the monster Dreaming dreams of fame.
Are you captured by your remorse introverted by your shame?
Do you hear the cries of hunger do you hear them call your name?

(chorus)

Are you stuck inside an office working 9 to 5 a day?
Results don't seem to matter, no you just collect your pay.
Are you ridden by the rule books and not what the people say?
Do you hear their cries for justice but just send them on their way?

(chorus)

Since that time countless people have asked me to record this song. I have always passed it off.

Just recently a man named Robert Wrigley, whose beautiful wife Bobbi Jo does ceremony with us, suggested to me to record the song. His two sons are a duo known as Double Trouble because they are twins and awesome singers and musicians. They also said I should record the song.

So now I am in the process of getting this song recorded with maybe a couple of others I have written. Wish me luck.

Now… back to the story…

Dreams guide, heal, show, calm and influence me. Here is another awesome dream that was given to me.

One winter's night with a blizzard blowing, I listened to the wind moaning as I lay in the little room that was my home. I was lonely and I was wondering where I was going. I was slowly building a new world that did not involve drugs and alcohol, but did it have to include this hole in my stomach that wind whistled through? I fell asleep and again my ancestors came and got me. They took me flying across a great desert.

It was summer time. In the distance I could make out mountains in the shimmering haze. We landed when we had crossed the space that was the desert. We were on the edge of a camp that had many teepees. As we entered the camp there were a number of people preparing for a ceremony. They were setting up their pipes in a large arc facing the teepee village. I watched and noticed some of these people with pipes had beer caps fall out of their blankets and bundles as well as beer labels on their person or on the stems of their pipes!

An Elder approached me and asked, "Are you going to set up your pipe and join us?" I was not a pipe carrier at the time but in this dream I said, "I cannot take my pipe out because I do not trust these people who are here Grandfather. I wish to join you but I cannot." He looked at me and said, "Well that is your choice. If you change your mind you can join us." I looked at him and said, "I will not change my mind Grandfather, as long as I see what some of these people are carrying and how this makes me feel." I was crying when I told him this. With that my ancestors told me it was time to go. Again we went flying across the desert. As we got about to the centre of this vast space my ancestors landed and told me to wait there. They said someone wanted to come and see me and then my guides left me standing there in the middle of the desert.

As I stood there looking into the shimmering haze the heat was producing, I saw that something was coming. As I watched I saw an image gliding along the floor of the desert and as it got closer I saw it was a person on a horse. As I watched the vison became clear to me and I saw a woman on a pinto horse galloping towards my position. I stood and stared and she approached… fast. She galloped up to me and reined in her horse just as she was directly in front of me. Her horse danced around in a circle and then walked completely around me. The woman on the horse looked at me as the horse danced around and said, "Sometimes we are lonely… do not miss me. I am on my way." After she said this she galloped away into the haze.

My ancestors once again took me flying back to where we started our journey. They set me down and looked at me and said, "What just happened to you?" I repeated all that had happened as I remembered it and they looked at each other and said, "Good… now he will remember." With that I woke up with the dream playing in High Definition in my memory. I felt happy… not lonely any longer. Eventually the dream's events were stored in my memory vaults. I continued on.

About seven years go by and I have graduated university with my B.Ed. and am working at a culturally based school in the city. I have met my wife and we have three children. One morning I am walking down the stairs of our house coming to breakfast and I hear my wife talking about these dreams she keeps having about pinto horses and how she is riding one across a desert! The dream I had about the lady and the horse comes galloping out of my memory banks and I begin to laugh. I rush downstairs and say to my wife, "It was you!"

I never told anyone about this dream and now I tell my wife and kids. I tell the kids, "Your mother told me she was on her way to me when I was lonely, sick, and homeless. She told me… 'Sometimes we are lonely. Do not miss me I am on my way.' I guess I never missed her or passed her by because she is here and because of her YOU are here with us!"

Other dreams have come to me to give me guidance, hope, solace, instruction, and purpose. I have dreamed dreams and they have come to be. I have found when people come and ask, with tobacco in hand, the spirits speak through the tobacco and I can interpret the meaning of their dreams if they so wish.

I know that without the guidance and hope given through dreams I most likely would have returned to the streets and, from what I have seen of others who have done so, died.

IV

In the Spring of 1987 I was called to speak with our Elders Antoine and Emma Sand and my teacher William Dumas. I was wondering what I had done now? I was worried about what the meeting was about. This meeting happened to be right after a sweat lodge; I thought I had done something to offend the Elders and I was going to get put in my place by my mentor in front of our Elders. When I arrived I was told to sit at the end of the table and listen.

My mentor spoke to our Elders about what he wished to do. He said he was instructed by his pawakanuk (spirit-guides) to present me with a pipe… I was ready. The Elders listened and when he was done they agreed. I was in total shock! I could not accept this honor. I was not ready. I was not deserving. The three of them looked at me and said I had no say in the matter. The gift and responsibility was given in a sacred manner and it did not matter if I thought I did not deserve this or that I was not ready. The decision had been made. I was to fast and find what the pipe looked like. When this was done I would approach a pipe maker and have the pipe made. I would bring the pipe to a ceremony when it was completed and it would be sanctioned by the Elders and my teacher.

I fasted for four days and during that time I dreamed of a pipe sitting in clear water. In the dream I looked at it and as I did so, I looked up and across the stream the pipe was sitting in, and saw my teacher was standing there. I had found the pipe. The pipe was made for me by my friend Mino Mikan (Good Wolf/Luke Arquette) and I presented it to my Elders. I was told this:

"You accept this pipe... *ospawakan*... and take it into your family and home and look after it as you have been taken in by others. This pipe is like an orphan and you must treat it as such. This pipe is like the child of another and you must accept it as your own. If you cannot do this then you will have a hard time doing the work that has to be done."

I thought, "I can do this!" If this is so then I can take people into our home with the support and help from my wife. Our home can be a safe haven for those who need some support for a while. I thought of my kind, beautiful partner, Connie. When we met she had a little boy from another relationship and my ego had a hard time dealing with this. I fasted, I went to ceremony, I prayed as to what I could do. What was told to me was... If you love this woman... then you must love this child. She Ba Gizi Koo Inini... Colin... Under-The-Sky-Man is my son. He goes with us to ceremony and is accepted by all as our son. I have carried this pipe since that time and have given it away two times and both times it has returned to me.

In the year 1988 I was given responsibility of our ceremonial camp as my teacher went to work up north. After this the camp did not last too long because of lack of leadership. I did not have the people skills to calm conflicts of ideologies and egos that began to emerge. The camp split with bad feelings and hurt feelings. We learned, we grew, we moved on.

During this time I started to make ceremonial pipes under the guidance of my brother in spirit, Mino Mikan (Luke Arquette). One night I had a very clear dream. In the dream I was once again walking by a clear stream of water and again the world was colored in a golden hue. As I walked I looked into the water. I saw four pipes sitting there. Each pipe was on its own pipe stand at the bottom of the stream, visible in the clear water. There were two black pipes and two red pipes connected to their stems. I awoke from that dream on a clear summer morning and immediately began to make the pipes I saw in the dream. I completed the bowls in two days and it took another week to get the stems. Once they were completed, the urge I had to make them left me and I wrapped them up and let them sit. About two months later I get an early morning phone call from my teacher who is now in northern Manitoba. He says he needs four pipes… two red and two black. I say OK. I begin to describe the pipes he is looking for and he says yes these are the ones! How did I know? I tell him the pipes are waiting right here and have been waiting for two months. The pipes journeyed up north.

There are many instances where my dreams have guided me, giving me hope, forgiveness, peace, closure, and direction. Never more so than when I was instructed to take on the responsibility of Sun Dance.

I sun danced with the Anishinabe of Ditibineya-ziibiing—the Rolling River Place (Rolling River). I have danced with the Anishinabe of Sagkeeng—at the mouth of the river, at a place called Strong Earth Woman Lodge. I have danced with my people the Ininewuk, at Spruce Woods Sun Dance, and I have sun danced with the Dakota Oyate at Birds Hill Provincial Park.

I have assisted and learned at Little Pine Sun Dance, Saskatchewan; Goodfish Lake Sun Dance, Saskatchewan; Saddle Lake Sun Dance, Alberta; Selkirk Sun Dance, Sagkeeng-South Shore Sun Dance; Sagkeeng-North Shore Sun dance in Manitoba; Fools Crow Sun Dance, South Dakota; Gordons' Sun Dance, Saskatchewan; and Matheson Sun Dance in northern Ontario.

While dancing at Birds Hill I began to have dreams about eagles, whirlwinds, lightening, turtles, buffalo, horses, wolves, bears, moose, and trees. These dreams were vivid and reoccurring. These dreams were also scary… so I did what I always did when I lived on the street and too much responsibility was expected of me… I ran away.

I ran from these dreams but they always found me. I ran for four years until one day I could not run no more.

One day I was coming back from a sweat lodge in St. Norbert and I passed a new building that had recently opened on Main Street and Higgins Avenue called The Circle of Life Thunderbird House. The building had been getting a lot of coverage in the First Nations community of Winnipeg because it had a sweat lodge within its confines and there was a lot of opinion about this. People were saying the sweat lodge should not be there because it was on a site previously housed by bars and hotels.

As I drove by I noticed smoke coming from the chimney where the sweat lodge was located. I decided I needed to investigate. I drove into the parking lot, parked and went inside of the gate to where the sweat lodge was. When I got there, the fire was burning in the firepit/place and an Elder and woman were preparing the area for a sweat lodge. As I introduced myself I asked if any help was needed. This is how I met Elder Don Cardinal and his wife Allison.

I worked for and with this man for six years and during that time he taught me responsibility. He taught me that we learn as we go and at some point we have to apply what we learned to assist others. One day Don said to me, "What are you going to do? I am not going to be around forever." It was then I realized work had to be done. We cannot sit back and expect the Elders to do what has to be done. We as *Oscapewis* helpers, must begin to lighten the burden of our Elders. We must step up.

I told him about these reoccurring dreams I had been experiencing and he told me there is something you have to do… you know what it is.

I knew then I needed to begin the preparations for a Sun Dance lodge. I was given four years to prepare. During this time I was shown how the lodge would look. How *Okimaw Atik* (Sacred Tree) would look. How the sacred pipes would look. What main colors to use. Why the dancers would dance the way they would dance and when they would dance.

The dreams I had were very vivid. In one particular reoccurring dream, *Pinesewuk* (the Thunderbirds) were talking to me. I dreamed my wife Muskwa Achak Iskwew (Bear Spirit Woman) and I were standing in a big open area and lightning was flashing all around. It was exploding everywhere and things were being destroyed all around us. A large bolt of lightning struck us. I saw it coming so I grabbed my wife and held her tight and waited for the shock that would surely come. Nothing happened! As I opened my eyes I saw the lightning bolt had passed right through us and struck a fence post that was behind us and the storm continued on its way.

Another dream had us again standing in a large open area and the thunderbirds were singing in the sky. Thunder rolled. As we stood there and watched, a funnel cloud came dancing across the field ripping up everything in its path. Once again the funnel came towards us and again I grabbed my wife and held her close… closed my eyes and waited. Again nothing happened. I opened my eyes to see the funnel go right through us and continue on its way.

I had recurring dreams where the eagles, wolves, turtles, buffalo, bears, thunderbirds, and moose came and spoke with me. I awoke from these dreams with purpose, hope, confidence, and commitment. I could do this.

It was at this time that another Elder was put in my path. Once again I was doubting myself, my instructions and worrying about what people would say about my plan to hold a Sun Dance. I was helping at a youth gathering at a school in the city when I came across an Elder named Noah Cardinal. Noah is from Saddle Lake,

Alberta and was talking at the gathering. I met him sitting out-side the venue relaxing. I introduced myself and he began talking with me. I did not tell him about the doubt I was going through or all the questions I had about the Sun Dance, yet as I listened to what he said I felt so much better. As I walked away I realized he had answered all the questions I had concerning my doubt and all my worries!

As it turned out the Elder Noah Cardinal was/is the teacher of three friends of mine… brothers. I ran into Noah again at a sweat lodge my friend/brother Dr. Michael Hart was holding. Once again I came to the lodge plagued with doubt and questions and once again without prompting Noah addressed all the questions and doubt I had. Once again I left the ceremony feeling like I could go through with doing the work.

When the summer of the Sun Dance arrived I was amazed at the awesome people that rallied to the dream of Sun Dance and the statement that Don Cardinal said to me, that "IF IT WAS MEANT TO BE… IT WILL HAPPEN"… echoed in my mind. The Oscapewis worked hard to get all ready for the big event. We would dance for four days under the full moon of August. We were to dance under the full moon during this time and the women… *Iskwewuk*, would lead us and women would sing for us… to honor Tipiskawi Pisim and Iskwewuk.

The days leading up to the ceremony we had people fasting at the sacred tree and were doing sweat lodges. One day a huge thunderstorm blew up as we finished a lodge. The trees were bend-ing in the wind as rain and hail fell. Branches were breaking off trees and landing all around us. As I watched, one pledger crawled under a table to protect himself from the falling branches. I asked another to place tobacco in the fire and pray. As he did this, a huge branch from a large tree landed with a thud, right between us. Everyone looked at me and I knew right then I needed to remain calm. I prayed and as Dwight put the tobacco in the fire it calmed right down. He has some powerful prayers that Dwight! I heard later that some ladies that left the sweat lodge just as the storm began were blown off the highway by the wind! No one was hurt!

We were told that during this time of preparation things would work out as was meant to be. One day prior to the event unfolding, a group of the ladies led by my wife, went to offer *wipinason*… prayer flags to Okimaw Atik… the selected Sun Dance tree. My wife said as they entered the scared grounds and looked at Okimaw Atik… the ground was covered with *Pipichiwuk*… robins. Hundreds of them! The birds took flight as the women approached. The women were in awe… scared… crying… laughing.

Pipichiw the robin is a very sacred bird to the People. The sun dancers blow a whistle when dancing called *pipitachikan*… robins' whistle. The sound that is produced with the short trill of these whistles… repeatedly… is said to emulate the sound the baby robins make when calling for food from their parents. Thus the sun dancers are calling for Spiritual Food for the Creator.

Pinesisuk… the Birds look after us that four year cycle. Cranes, eagles, robins, and thunderbirds cared for us as we danced the dance of life.

My brothers said that our dad was there dancing with us.

My father's nickname was Pinesis… Bird. He drank himself to death with the pain he carried.

We were called to The Pas to see him before he journeyed home. He waited till we got there. Our son Mistawasis was just six months old at the time. We stood by the hospital bed, my wife holding Mistawasis in her arms and my father lifted his frail arm to Mistawasis and smiled… then he passed on.

Pinesisuk (the Birds) watched over us during that first Sun Dance cycle.

Another time during the dance lightning struck the sacred tree! We saw the lightning strike from the west side and leave the tree through the east side… we continued dancing in the driving rain.

It was awesome!! Pinesewuk (the Thunderbirds) HAD COME and BLESSED US!

All the sun dancers who pledged to dance with us stated as they entered the dance area, "I have dreamt of this place... or... I have been here before."

At the beginning of the dance I was very worried for the dancers as well as for the lack of Elders we had with us. It was prior to the starting of the first year of the Two Suns Prayer Lodge Sun Dance that our Elder Don Cardinal returned to the Spirit World. I was scared... I was scared I would hurt someone or that what we were doing would not be acknowledged by the people. When we began to dance, early morning, that first day I looked up from the business at hand and saw about ten Elders seated in a line, looking at us! I was amazed and began to cry. Those Elders stayed with us the whole time through the four years of our commitment. During that time they requested that we bring the dance up to their community, Nisichawayasihk... Nelson House.

I thank Jackie and Felix Walker for their commitment and allowing my family and I into their lives.

We were also instructed by one of the Elders to use a bull moose skull in the ceremony as well. Some Sun Dances honour the buffalo by dragging the buffalo skulls as part of the commitment and sacrifice as well as to acknowledge the gifts the buffalo gives us. The Elder from Nisichawayasihk stated that up in the north country it was not the buffalo that fed our people... it was *moosa*... the moose. The Elder, old man Josh (Linklater) worked with two of our dancers, Jackie and Felix Walker, and they were the ones that brought the Elders down from Nisichawayasihk. Felix Walker went out hunting with old man Josh and this one time they had shot a bull moose and Felix had taken the skull home. He usually gave the skulls away but for some reason, he said, he decided to keep this particular skull. It sat on the roof of his shed for three years until one day we were looking at it and he asked if he should bring it down to the dance with them. I said sure and left it at that. As it turned out this was the moose skull we dragged to honor the sacrifice made by moosa. It felt very right to do. We also use the tongue and heart to put under Okimaw Atik to acknowledge the sacrifice and gifts *Mistikuk*… the tree nation, gives us.

In the second cycle of the Sun Dance we moved up to Grand Rapids, Manitoba where our relatives, the Cook Family, hosted the ceremony.

We were blessed with visits from Muskwa… the Bear during this dance. Our relative watched over us. One time our brother Muskwa came right though the Sun Dance grounds and a Conservation Officer (CO) had to be called BECAUSE THE BEAR WOULD NOT LEAVE!

When the CO got there, he sprayed the bear with bear mace to try to make it leave the area. The mace had little effect on the bear. The story that came out of this event was that when the bear got sprayed… the bear reared up on its hind legs and began making motions with its front paws… like it was smudging! Looked like he was washing himself with the fog!

Very hilarious… funny. Our brother Muskwa danced with us!!

The third cycle of the Travelling Sun Dance… as this Sun Dance was dubbed, found us in Nisichawaysihk… Nelson House, northern Manitoba.

This cycle we were blessed with our older brother… *Nistis Mikan*… the Wolf. Mikan came and watched over us…

One point Mikan was spotted on the side of the road as some ladies were leaving the sun dance grounds.

The ladies, being from the south, were very excited to be able to see a wolf and began taking a video on their phone to add to Facebook. As they took their video they did not realize how close the wolf had approached their car.

When they saw the wolf right up close to their car window they panicked! The lady taking the video began to roll up the window… but did not pull her phone back in! The phone fell to the ground where the wolf picked it up in its mouth and took off for the edge of the forest!

The ladies yelled and screamed at the wolf but the wolf just looked at them. *I have your phone* it seemed to be saying. It was not until the ladies took out their hand drum and started pounding on it did the wolf get startled and run into the bush… dropping the phone in the process.

Needless to say the ladies recovered their phone and did they have a story to tell!

Of course First Nations humour steps in and the story was that the wolf was taking selfies.

Next cycle we are moving "The Travelling Sun Dance" to Grassy Narrows Ojibway Nation in northern Ontario.

Wonder what awaits us there?

The Travelling Sun Dance.

My family has ever been present and supported all that is done with our community.

I have been given the responsibility to watch over, love, feed, teach, worry about, and look after four awesome children and a wife that does all the aforementioned for me! We as a family also take care of my mother's older sister who my children refer to as Granny. She is our matriarch and when Connie and I were attending university, Granny watched over the kids. Now it is their turn to watch over their Granny.

Other responsibilities include to carry a sacred pipe, carry the sweat lodge, interpret dreams, find sacred names, pawakanuk (spirit guides) and colours, officiate weddings and funerals, conduct Fast Camps and carry the sacred Sun Dance. I am a sun dancer of twenty years and a teacher and a student until the end of my days.

A few years ago I had an operation to remove a cancer tumor that was in my small intestine and part of my colon. The operation went well but I ended up with some kind of infection and had a very high fever. In the hospital I had a vision of young people standing up against oppression. Some wanted violence. Some others asked me to speak with these youth. The youth had guns, knives, metal bars. They were angry. I spoke with them and as I was doing so something happened in another area and spread very fast. Violence erupted all around. As I stood there trying to calm people down I was struck by a sword or heavy bar. I lay on the ground crying for the youth. I faded away.

When I awoke from that dream it was in a dim twilight and there were people standing at the edge of my bed whispering. I heard one say, "he is not out of the woods yet… we will keep an eye on him." As I listened to what they were whispering I noticed I was drenched in sweat, as well as my blankets! I was wet!! I slowly faded back into oblivion. Went to sleep again and when I awoke it was into a bright sunny day. I had returned again! I continue on.

I had been ruled by my physical world when I was a child. I depended on those around me for survival and this I was given. As a youth I was ruled by emotion from one extreme to another. I was a calm young person and an angry young man. I saw the black or white… truth or fiction, good or bad. Emotion carried me away for during this time I had none of the checks and balances that family provides. I explored, experimented, and no one could say otherwise. As an adult I was ruled by the quest for knowledge, the building of my mental capacity… reason. Now as I enter the spiritual realm I try to see things differently, from various perspectives. I try not to be a reactionary. I try to remain calm and quiet.

The physical, emotional, mental, and spiritual stages of life… newo (four). Ininew… the four bodied, the mixing of four (water, earth, air, and fire). I have tried and continue to try to be a good man. A person who my wife, my children, my friends, and my people can speak about and smile.

I have also worked as a security guard, mucker for Inco mining Company, carpenter, counsellor at a half-way house, in a treatment centre, in a family violence men-who-batter program, as a youth worker, cultural advisor, and truck driver. I hold two degrees in education from the University of Manitoba, taught for twenty years at the early, middle, high school levels, and various university courses as a guest speaker.

Presently, my glorified title is Science Facilitator with the Manitoba First Nations Education Resource Centre (MFNERC) and my mandate is to put First Nations perspectives into the sciences. As part of this responsibility I have been given the job of finding and re-establishing the traditional knowledge of my people as it pertains to Achakosuk... the stars. This is one awesome job and I have the opportunity to meet holy, reverent, calm, awe-inspiring people and travel to places of learning and teaching to assist in the broadening of views pertaining to First Nations thought, perspective, philosophy, ideology, and cosmology. I have been shown constellations, stars, and given stories, ideas, philosophies, hypothesis, mythologies, songs, and have been told how all this connects us to the larger reality... miswa... all that is.

As I do this work I have slowly come to realize the immensity of knowledge contained within the philosophy and world view of my people. The worldview of my people is in direct contrast to what Western philosophy and ideology has taught me to believe, the idea that man is the apex of knowledge and holds dominion over all. I was told this in grade school, in church, in university, on television, and in books. The teachings of my people tell me we are but a small part of miswa... all that is. We cannot begin to comprehend the enormity of creation. The teachings tell us we are reliant on and need all things in order to survive and thus must remain humble. This is the perspective our Elders pass along to us:

> Kitim ma ki in...we are pitiful. If one morning we awoke and something had happened to mistikuk, the trees, what would

happen to us? The answer is we would perish. If one morning we awoke and something had happened to assiniuk, the rocks, what would happen to us? Again the answer is we would perish. If one morning we awoke and something had happened to muncoosuk, the insects, what would happen to us? Of course the answer is we would perish. Now, if one morning the w·orld woke up and something had happened to the humans, what would happen to the world? NOTHING!! The world would continue on.

The point being made is… THIS is how important we are to the world. We need all things in order to survive and all things do not need us. We must never forget this.

The teachings go on to state miswa is incomprehensible because Creator is incomprehensible. Limitless… What has limits is the human mind because it is anchored to a physical world… but residing within us is Achak… spirit, and this has been touched by Creator. We are told Achak does not end because our Creator does not end. As Creator creates miswa, we understand it is ever changing… never ending… always more being made. This is because Creator is ever thinking and as Creator thinks… creation is made. What that means is that right now, at this very instant, creation is still in the process of unfolding…

We are told Creator can never end… therefore the part of Creation that has been touched by our Creator can never end. For the Ininewuk this part is Achak… the spirit. Our spirit is energy… light, and continues on its endless journey when our visit to this physical world is done. Achakuk… spirits/energy travel miswa, the cosmos, searching for new understanding, teachings, learnings, experiences. Through beings of spirit/energy identified as *kisikookuk* … beings of light, we come to this place we call Aski… earth.

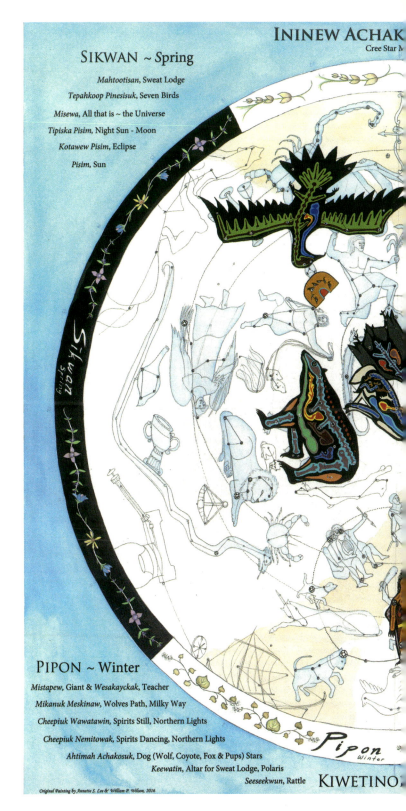

INININEW ACHAK

Cree Star M

SIKWAN ~ Spring

Mahtootisan, Sweat Lodge

Tepahkoop Pinesisuk, Seven Birds

Misewa, All that is ~ the Universe

Tipiska Pisim, Night Sun - Moon

Kotawew Pisim, Eclipse

Pisim, Sun

PIPON ~ Winter

Mistapew, Giant & *Wesakayckak*, Teacher

Mikanuk Meskinaw, Wolves Path, Milky Way

Cheepiuk Wawatawin, Spirits Still, Northern Lights

Cheepiuk Nemitowak, Spirits Dancing, Northern Lights

Ahtimah Achakosuk, Dog (Wolf, Coyote, Fox & Pups) Stars

Keewatin, Altar for Sweat Lodge, Polaris

Seeseekwun, Rattle

KIWETINO

Original Painting by Annette S. Lee & William P. Wilson, 2016

...OS MASINIKAN
...ap-Book

NIPIN ~ Summer

Nipin Pinisew, Summer Thunderbird
Niska/Niska Meskinaw, Goose/Goose's Path
Nipinesis Meskinaw, Summer Birds' Path
Achak Sipi, Spirit River, Milky Way
Achakosuk, Stars
Askiy, Earth

TAKWAKIN ~ Fall

Kokominakasis, Grandmother Spider
Pakone Kisik, Hole in the Sky, Pleiades
Achakos Ahkoop, Star Blanket, Pleiades
Mahtootisan Assiniuk, Sweating Stones, Pleiades

Pipon Pinisew, Winter Thunderbird
Mista Muskwa, Bear & *Ochek*, Fisher
Makinak, Turtle

...HK ~ North

ININEW ACHAKOSUK MASINIKAN (Cree Star Map)

Map Created By Wilfred Buck, Manitoba First Nations Education Resource Centre Inc. © 2016

Our mythology tells us the first *Kisikook* (being of light) who ventured here was a being called Achakos Iskwew... Star Woman, and through her and ALL our female relatives, we come to this place we call Aski to visit. This refers back to the term pimatisiwin.

Looking into the night sky we identify a group of stars... achakosuk, as Pakone Kisik... the hole-in-the-sky. To mainstream astronomy these are the Pleiades, the Seven Sisters. The mythology says Achakos Iskwew (star woman) was roaming the cosmos and came across Pakone Kisik (the hole-in-the-sky). As she looked through this hole in the sky she saw this planet, Aski (earth). After observing this place for a while, Achakos Iskwew decided she wanted to go to Aski to learn, experience, teach, and live. She found she could not just go through Pakone Kisik and do and experience the things she wanted to experience. She needed assistance.

She approached another of her kind called Kokominakasis—Grandmother Spider, who sat on (what we now call *Achak Sipi*—river of spirits) the Milky Way. Kokominakasis held all access to Aski and knew there were certain conditions that needed to be addressed in order to experience Aski. Kokominakasis told Achakos Iskwew, "if you want to visit this place there are three conditions that you must agree to.

One: you must take a physical form when you go to Aski.

Second: you cannot stay there forever. At some point you have to leave.

Third: You must bring a gift that will comfort and remind all who come after you... where it is they come from and where it is they return to when their visit here is done."

Achakos Iskwew agreed to these conditions and Kokominakasis sent a single strand of webbing through the-hole-in-the-sky and lowered Achakos Iskwew to Aski.

The gift Achakos Iskwew bought was the star blanket which had seven points to represent the seven visible stars of the Pleiades.

The other condition agreed upon was that she take a physical form and this form she took was us, Ininew… or if you prefer humans.

The length of time Achakos Iskwew would visit was a lifetime. For a being of energy how long is long, so a lifetime would suffice.

This is how we all come to this place, through Pakone Kisik, to begin our journey of a physical life. The single strand of webbing we are lowered down with is called *mitisai* or *mitisayapiy*—umbilical cord. This term is the root word for the term my people have for life, Pimatisiwin.

We are star people and when we leave "this mortal coil" we continue our never ending journey amongst the cosmos.

This teaching holds tremendous impact for those who hear it and comprehend what it truly represents. A totally different perspective of life after death, spirit, man as the image of god…

This teaching relates a totally 'alien' worldview which puts into question the Catholic Church and Christian doctrine and dogma. Here is presented a totally different perspective on our place in the cosmos… our thoughts of an after-life… religious hierarchy… a physical building of a church… burials… heaven and hell…

I could imagine the heads of the holy church, who controlled every aspect of life in their society… hearing about kisikookuk (beings of energy/light), alternate realities, no heaven or hell! This would have threatened their worldview and whole power structure!

Something to think about…

On a side note… I was telling this story inside the portable planetarium one day. The dome was crowded with people coming to listen to the mythologies. One particular participant, a certain Leonard Sumner, a local Indigenous recording artist and speaker, was sitting near me inside the planetarium. As I was sharing the story of Achakos Iskwew (Star Woman) he silently bent nearer to me and whispered, "What is the name you are calling this woman? Pachakos ekwew??"

At this point I crack up laughing. Leonard is Anishinabe and *pachakos* means "ugly" so he was jokingly asking if I am calling this sacred being "ugly woman." No I reassured the audience, I am not saying pachakos... but Achakos. Lol...

If you look on the map (*Ininew Achakosuk Masinikan*) you can see the Milky Way running across the painting highlighted in a light tan color. The Roman Greek constellations are laid out in the background as a frame of reference for people used to seeing these in a certain position in the sky. In the foreground are some of the *Ininewuk Achakosuk* (Cree Stars).

Kokominakasis, Grandmother Spider sits on, in this case, Pawamiuk Sipi... River of Dreams, the Milky Way. If you are familiar with the classic eighty eight Greek Roman constellations then you can identify Queen Cassiopeia sitting on her throne... the famous "w" in the sky.

The inner star of the "w" is the head of Kokominakasis while the left and right sides of the "w" make up the front legs of Grandmother Spider. She is in charge of the doorways to the other realities. She was also responsible for managing our dreams... our glimpses of other possibilities. She is the one that gifted us with the Dreamcatcher which filtered our dreams for us.

On another side note: in the portable planetarium I take around to the various schools telling the children and youth about their histories and knowledge present in the night sky, I introduce the famous "w" in the sky... sitting right inside the Milky Way. I say to them, "you see that 'w' in the Milky way right there? THAT is a very important constellation. That 'w' stands for WILFRED!!"

Very important! From there I go on to tell them about the ININEW ACHAKOSUK... a First Nation perspective. Their Perspective.

One of the most interesting things for me about these sacred stories is that the constellations are interchangeable, depending on what knowledge is being put forward. This concept is very frustrating for academics who want to see one perspective from one constellation... a rigid, static system. For my people, this is not the case. The constellations move and flow in relation to where we are at this time and space in our own concept of reality.

If we take Pakone Kisik (the-hole-in-the-sky/Pleiades) as an example, we get many perspectives.

In one perspective, we see this group of stars as a place of origin, a doorway to multiple realities... the-hole-in-the-sky. Where we come from and where we go back to.

In another one we see Matootisan Koonaci Iskitiw, the Sweat Lodge sacred fire, where spirits are readied for release. When we leave this reality... a sacred fire is lit and burns for four days and we make our journey back home to the cosmos.

In yet another perspective the hole-in-the-sky is represented by Acakos Ahakoop, the star blanket, reminding us where it is we come from and where we return to as well. This reminder also makes us remember all our female relatives that have come before us... we owe them our existence at this space and time.

These interchanging, interacting, interconnected perspectives are common practice when talking about achakosuk... many perspectives, many meanings but all interconnected.

There are other realities out there. When Pakone Kisik is mentioned, what is accepted as fact is the concept of alternate universes/realities. They are talking about a spatial anomaly... a worm hole. At this time and place where our reality exists is but one of many possibilities. They say Pawamiuk, our dreams are another example of other realities. Dreams give us glimpses into possibilities and possibilities are endless. Another example of this phenomena are visions. We venture into the multiverse of alternate realities, particle and quantum theory... energy and possibilities.

Another thought that is mind bending is that every culture... every culture on the face of the earth had/has their own sacred stories, perspectives, and relationships to the star world.

EVERY CULTURE had/has a depth of knowledge and an intellectual capacity. This is a universal occurrence. It did not just happen at random to one specific culture… just as where we now reside is only one of many possibilities out there. Earth is but one of billions of planets… in one galaxy… surrounded by billions of galaxies… occupying space in multiple realities…

I have come to see that the depth of knowledge possessed by a people is not central to the civilization, environment, weather patterns, social, or population distributions. All people have delved into the questions of who we are, how we got here, why things are the way they are, what happens when I do this, and what possibilities are out there?

It is interesting to note here that Indigenous People from all over the world see Pakone Kisik as a very Holy place…

Our Elders have told us that the sky is full of stories, guides, teachings, and answers to some why or how questions. If all the sky knowledge of the people were available, we would have star maps that would cover the whole sky. But, sad to say… this is not the case. Due to the collapse of our nations and world a large amount of knowledge has been lost. One Elder, Don Cardinal of Sucker Creek Cree First Nation in northern Alberta, put it this way:

> With the arrival of the Europeans and first contact was made… the collapse of our world began. When the people met each other on those far away islands in the Gulf of Mexico, the seeds of our destruction were sown. Prior to the arrival of the Europeans there were vast trade routes that were established throughout the Americas. Trade goods went in all directions.
>
> An example of this is… in 1969 (or thereabouts) on the south bank of the Saskatchewan River and mouth of the Opaskwayac River in the town of The Pas, Manitoba, while digging, excavating for basements, a burial site was uncovered.

In that burial chamber was a female wrapped in a cedar cloak. Anthropologists from the University of Minnesota came and examined the find. They took the artifacts back to the university where some of the items were analyzed.

The cedar cloak had originated on the west coast, what is now British Columbia. She was colored in red ochre that originated in the Dakotas and had in her possession an atlatl of which the dart tips were made of flint chert that came from the Great Lakes area. She also had a shell which originated in the Gulf Of Mexico. Thus to look at the items of cedar cloak, red ochre, chert and sea shell and you can see these were trade goods and that the trade was not only from the immediate local vicinity but spanned a wide area.

When the Europeans and Caribes made contact with each other, trade goods exchanged hands and these roaming Traders took these trade good far and wide. Unbeknownst to the traders, other things were being traded also. Diseases were carried inland along the vast trade routes and it is estimated up to eighty per cent of the nations of North, Central and South America were decimated by these newly introduced diseases.

To put it into terms we can comprehend, look at it like this. A village of one hundred individuals exists somewhere in the Americas. All the knowledge of the people reside within the experience, philosophy, understandings, and imagination of the various Elders, orators, leaders, thinkers, hunters, medicine people, dreamers, etc…etc. All the people who comprise a total community from day to day affairs to philosophical schools of thought… the total knowledge base of your people. One morning the village wakes up and eighty individuals are gone.

Think of it in terms of:

Each individual holds knowledge… or we will say a single word to a song. In your particular village of one hundred individuals the total knowledge base is held in a song comprising of one hundred words of which everyone knows one word. Together the whole knowledge base of your people is revealed when everyone gets together and sings their part of the whole one hundred word song. One morning you wake up and eighty people have ceased to exist. Twenty percent of the knowledge base remained. Eighty percent of the total knowledge of the people has been wiped out. With this twenty percent of total knowledge remaining, the people needed to survive, relearn, repopulate, recover lost knowledge, and move forward. This does not only happen once but happens wherever Europeans encounter the original peoples of the Americas.

This is what has happened to our people.

But... I was told... this knowledge that was lost can be recovered. Through education, Sun Dance, fasting, Sweat Lodge, dreams, visions, sharing, and meditation we can again regain the vast knowledge of our ancestors. It is happening now... because how the original knowledge was acquired was through travelling, ceremony, dreams, sharing...

One morning I happened to be arriving at the First Nations University of Canada in Regina, Saskatchewan and was invited to sit in on a pipe ceremony. After the ceremony I got to speaking with one of the Elders that was present, Elder Ken Goodwill.

He had asked where it is I originated from and what I was doing on my visit to the university. I said I was there to inquire about any star knowledge that people were willing to share. I explained what I was doing and how that knowledge that was shared would be given to our young people through the medium of the portable planetarium. He liked that idea and asked if I could share a story I knew with him as an example of what I was trying to do. I shared the Mista Muskwa (the Great Bear or Big Bear) story with him. After hearing the story, he was very interested in hearing more.

He said his people, the Dakota as well as the relatives down south, Lakota, and relatives to the west, Nakota who made up the

Sioux Nation held a great deal of knowledge about the sky and the stars. He said his people had been lucky in that all the "star knowledge keepers" did not die when the sicknesses came across the land.

He then said that if you were to look into the night sky on a clear night, with your vision unobscured by anything on the horizon, all the stars you could see with the naked eye our people knew.

This was common sense because the people lived under that sky twenty-four-seven and observed it whenever they were out and the sky was clear. In the winter months, stories were told during the long winter nights to educate as well as occupy the minds of the young people who were confined to just the surrounding areas of the camp.

Every visible star and celestial object had a name or was attached to a constellation, had an important time frame, teaching, ceremony attached to it, songs and instructional methodologies.

Standing with Elder Ken Goodwill I was in awe at the vast knowledge our people possessed/possess.

He went on to say that with colonialism, assimilation policies, genocidal acts, diseases, and historical trauma that happened to our people, we lost anywhere from eighty to eighty-five percent of the knowledge base our people once held.

This only inspired me more to find as many constellations and stories of my people as well as other Indigenous star knowledge.

He is the one that said to me, "this forgotten knowledge can be relocated. There was a certain process as to how that knowledge came to our people. We have not stopped doing some of the things that brought that knowledge forward. We still have Sun Dance, Sweat Lodge, Fast Camps, Shaking Tents, dreams… songs, ceremony and prayers. We still travel and share with one another. This is how knowledge is gained. This is what our people did and this is what we can do. Nowadays we also have educational systems and technology to assist us in the work that has to be done."

The story of Mista Muskwa, the great bear, was told to me

by a knowledge keeper from the Edmonton area a long time ago. Maybe thirty-nine years, maybe a little longer? If I remember correctly his last name was Twinn. I was at a treatment centre called Poundmakers Lodge in St. Albert, Alberta trying to get sober. It was the second star story of my people I had heard up to that point. The first one was of Ocik… The Fisher and the Big Dipper told by Elder Murdo Scribe from Kinisiw Sipi (Norway House, Manitoba).

Since that time I have come across the story in various forms among the people of the Rocky Mountain House area in Alberta and the Mi'kmaq on the east coast. Of course these stories have their own variations which reflected the environment and fauna of the various areas.

The Big Dipper and Mista Muskwa story goes like this:

Kayas… long, long ago, before the coming of the Ininewuk, huge bears (Mista Muskwauk) roamed all over what is now known as the Americas. This story concerns one such being. This particular bear (Muskwa) was considered the biggest and strongest of all the bears. As the bear roamed it encountered all beings residing on these lands. It found that these beings were more than ready to offer it anything it wanted as a way of getting on its good side. As time went on, it got to a point that the bear expected great tribute when it arrived somewhere. When none was forthcoming the bear would get angry and threatening. The great being found that if it acted this way it could get anything it wanted and if it wasn't given, it could take it!

It got to a point that the bear just invaded camps and communities, destroying and killing just because it could. Years went by and all the beings lived in fear.

One day, a great meeting was called to see what should and could be done about this threat that hung over all living things. The bear heard of the meeting to be held but made no response. It was confident in the power it wielded and the fear that power produced.

Let them have their meeting, the great bear thought.

When the great meeting finally took place, after a long discussion it was decided that the bear would have to go. No amount of kindness, negotiation, friendliness, gifts, or submission appeased the bear. It did what it wanted to do and took what it wanted to take.

It came down to choosing how and who would be sent to do the dangerous deed. Seven small birds (Tepakoop Pin niss sis uk) were chosen. Seven of the best trackers, most skillful hunters and least noticeable birds were sent to do the job.

Among the seven hunters was a not too large, unassuming, plain greyish brown bird. Its name was Pipiciw... Robin. Now during this time of the world, the robin did not have a fine red breast or colorful blue egg. It was a plain, unassuming bird.

So this is where the various versions of the story change with what birds were sent, what time of year it was as well as how the bear now stands in the sky.

We will continue on with the story the way I first heard it mentioned on the northern plains.

The birds took up the hunt and tracked the great bear down. When the bears saw the hunters coming it laughed and thought, "what are THEY going to do against the might and power I possess?"

So the bear laughed and laughed as the birds came forth. When they got close the bear charged them and made loud and scary roars. The birds flew high into the trees always watching what the bear would do next. Of course the bear attacked the trees but the birds just flew out of reach.

After the bear had spent a large part of its energy on trying to scare the hunters, it got tired and went to go rest. Upon seeing this the seven hunters immediately attacked the bear... severely punishing it and giving it no rest. This pattern went on for a long time until finally the bear was totally exhausted. The birds were taking turns harassing the bear. They had divided themselves into two groups and took turns at their grim resolve… ever harassing and hurting the bear.

Finally, it got to a point where the bear was so desperate to find rest that it began running in an effort to find a place where the birds could not bother it. The chase was on!

We are told the chase took the participants around the earth four times. As the chase evolved it gathered speed and we are told that on the fourth time they had gone around the earth they came up over the horizon and flew into the sky they were moving so fast!

Now this happened to be around the months of September/October, or as it has come to be known autumn, as we now measure time, when they flew into the northern sky. As this took place the bear was near total exhaustion and turned to face his harassers.

When the bear did this, one bird, the robin, managed to get a shot at the bear that mortally wounded it. Mista Muskwa, mortally wounded, raged and raged in the northern sky. It began to bleed profusely. Blood was falling everywhere. We are told the bear bled so much it began to shake. It tried to get a clear vision as to where the hunters were and upon doing so shook… just like a dog shakes when it is wet. When Mista Muskwa shook all the blood it was bleeding flew off of its fur and fell to the earth. All that blood landed on all the broad leafed plants and painted them orange, rust colored, and red.

This is why the leaves change color in the autumn, because of the blood of that bear.

They go on to say that when the great bear shook the blood off of its body, droplets of blood hit the bird that had mortally wound it, right on the bird's breast. That bird was of course the plain, unassuming, greyish bird…Pipiciw, the robin.

So today Pipiciw wears a red breast, because of the blood of that bear.

The end of the story says Mista Muskwa was placed in the sky to remind us of how we should live and what power and unlimited power can do to a spirit. We should endeavor to be kind, humble, helpful and caring if we ever are put in a position of power and influence, or we might end up like that bear. It also tells us that even the most down trodden, small, weak beings can come together, work as a unit and bring down the most powerful of beings. THIS is Community.

The story goes on to say that, to ever remind us of these things, Mista Muskwa and Tepakoop Pin niss sis uk were placed in the northern sky. The seven birds would keep an eye on Mista Muskwa.

Lastly we are told that Pipiciw is the brightest star (*Alphecca*) in the Tepakoop Pin niss sis uk constellation (Corona Borealis). Pipiciw was further honored by being given a special egg... the color of the sky... sipikook... blue. On the egg were placed speckles to remind us of achakosuk (stars) in kitcikisik (big sky).

TEPAHKOOP PINISISUK

A further honor given Pipiciw was that the distinct cry the baby robins make when they are asking their parents for food is imitated by the sun dancers when they blow their whistles, which my people call *pipitacikun* (sound of the robin). Root word is pipiciw. The sun dancers cry for pity and ask Creator for "spiritual food"…

When we look at the big dipper we are reminded of Mista Muskwa. The end star of the handle on the dipper, Alkaid, is the head of the bear. Alcor and Mizar, the binary stars at the bend of the dipper's handle represent the neck of the bear (Mizar) and arrow the bear was mortally wounded with (Alcor). The bowl stars of the big dipper make up the body of the bear.

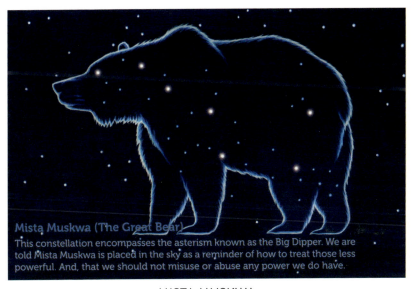

Mista Muskwa (The Great Bear)
This constellation encompasses the asterism known as the Big Dipper. We are told Mista Muskwa is placed in the sky as a reminder of how to treat those less powerful. And, that we should not misuse or abuse any power we do have.

MISTA MUSKWA

On a side note: I have noticed recently on some of the more up-to-date apps of the stars and constellation, Ursa Major no longer has a long tail reflecting a better understanding of Muskwa.

As time marches on, more knowledge has been placed on my path. The knowledge of my people held in the stars is what fascinates me and it is these stories that seem to find their way to my field of vision.

One such story that touches my spirit and heart is the story of how human beings acquired dogs. Mikan, the Wolf is my relative and has guided and protected me at various times on my journey from then to now.

This story concerns the stars identified as Ursa Minor… the little dipper. For my people, the name given these stars is Atima achakosuk… the dog stars.

Long ago during the time before Atim, the dog and Mistatim, the horse, came to the people, our relatives had no early warning system in place to warn of coming danger. They would set up night watchmen to watch over the camps as the people slept. But given that the night watchmen were human, they did human things. During their vigils they would get hungry and decide to go and eat. Or maybe they would eat too much and have to spend extra time relieving themselves. Maybe during their long vigils they would get bored and fall asleep or maybe during the lonely nights their honeys would sneak in to see them. It was during these times when the watchmen were otherwise occupied in something other than doing their job, raiders or dangerous animals got into camp, and chaos, destruction, and death occurred.

This pattern went on for years and years… but then one day our Nistis, Mikan (older brother the Wolf) felt sorry for us. Mikan decided to help us and sent two wolf cubs to live amongst the humans as warning systems, helpers, and friends.

Thus it came to pass that two wolf puppies came to live among the humans. Now, Mikan had two siblings, or *Nisimisuk* as my people call them. The siblings were Miscacakaniis, the Coyote and *Makisiw*, the Fox.

When the two siblings heard their Nistis… older brother sent puppies to live with the humans they said, "we also need to send puppies to live with our relatives, so our older brother will not be the only one who holds favor with our relatives the Ininew."

So it was that two coyote pups and two fox kits were also sent to live with the humans. Thus six puppies from the first three True Dogs (Wolf, Coyote, Fox) went to live with the humans. We are further told that there were actually twenty-four puppies that went to live with the humans in all the four directions of the world. Six to the North, Six to the East, Six to the South, and Six to the West.

From these puppies stem all the dogs of the world… twelve interbreeding pairs of three species…

Thus it was said that Creator was pleased and gave the humans dreams that identified the three first true dogs as the stars: Mikan (Polaris), Miscacakanis (Kochab), and Makisiw (Yildun), with the four stars in the bowl of the little dipper representing the four directions of the world to which the puppies were sent.

An interesting bit of information I came across while finding this awesome story was that the Cheyenne people had/have a military-style society called the Dog Soldiers or Dog Men.

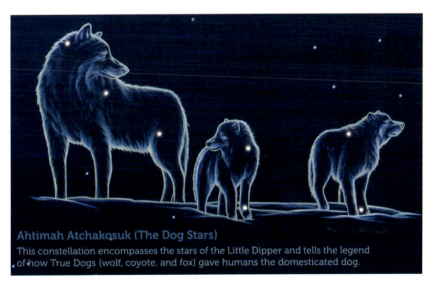

Ahtimah Atchakosuk (The Dog Stars)
This constellation encompasses the stars of the Little Dipper and tells the legend of how True Dogs (wolf, coyote, and fox) gave humans the domesticated dog.

ATIMA ACHAKOSUK*

It was said that one of these sacred warriors had a dream/vision about the little dipper and how if you observed it for any length of time you would see that it spins around the sky. It looked like the dipper was anchored to the sky because Polaris, the north star, is the last star in the handle.

Polaris looks like it remains stationary in the sky. My people call this the "star that stands still," *koskwawacikipiw acakos.*

So the dreamer saw the dipper spinning, anchored to the sky by the north star. In the dream he was told this was a representation of a dog, tied to a dog rope, which was staked to the ground. The dog guarded the camp and the radius the leash covered represented the protected territory of the dog.

The warrior was instructed to make a braided rope from the rawhide of dogs. He was told to make a wooden stake and wear both around his waist. If and when the camp was ever attacked and the people were in grave danger, the warrior would walk out before the enemy… tie one end of the rope to his waist and stake the other end of the rope to the ground. He would fight here… to the death if need be… to give the people a chance to retreat. Ideally there would be three or four of these "dog soldiers" staked to the ground forming a line of defense for the retreating people. The Dog Men or Dog Soldier Society…

At the moment I am working on identifying the stars of Mikwun Atik (Eagle Staff), the flag of my people. This constellation encompasses the stars of the big dipper, a star of Draco, two stars in Bootes, and two stars in Corona Borealis. I think there are thirteen stars in all that make up this huge constellation. I am in the process of finding an artist who will see and paint this vision.

During the last year things have happened to me that sort of galvanized my resolve to make some changes in my life.

I once again pick up the pen (computer), the guitar, the horse (iron horse… vehicle) and look in a new direction. As of December 31st, 2020 I will be officially retired from my present position as

Science Facilitator at MFNERC and will be setting out on a journey of exploration of possibilities.

I have another book deal, songs to be recorded, stories to be written, places to see, people to experience, dreams to dream, and life to live.

Just recently, I have been hospitalized with a heart attack and underwent triple bypass surgery. Once this was done I was told I needed a pace-maker to regulate the electrical impulses to my heart. So now I have had these things done and I am feeling great! Other than the fact that it has now been a year since my surgery and there are still two very stubborn holes in my chest that do not want to close. I am told this is most likely due to my diabetes and compromised immune system. My wife, Connie, changes the bandages and keeps everything clean of infection. She has been doing this for a year now and I am amazed with her dedication and care. I am a very lucky *napew* (man)! Hoe wah!!

One thing about these times is the anxiety that slowly creeps into my dreams and waking worries about my mother (Granny) who stays with us, my children (all four stay with us) and my beautiful wife who goes to work and interacts with high school students, teachers, administrative supports, and parents on a daily basis.

We try to be strong for our adult children but there are times we cry in silence, smudge, pray and lift our pipes to ask for guidance, calm, awareness, empathy, clarity, and hope. We all need something to believe in…

The dreams that are given me continue to guide, warn, heal, comfort, teach, and educate. I hold a deep respect for the Elders that have brought forward the vision, hope, ceremony, and philosophy of our people. Through disease, government legislation, racial oppression, systematic genocide, religious persecution, hate, shame, ridicule, and even death, they sent their message through the years to be heard by those still to come. *Chapanuk…* our great grand parents… our great grand children.

I will leave you with this the teaching of Chapan…

It is said we hold sway or influence over seven generations. The seven generations teaching can be seen through different perspectives… as seven generations from the signing of the treaties…. to seven generations from an oppressed mentality to… seven generations we hold sway or influence over.

As with all things, there are many versions of the seven generations teachings.

One tells us that when an important decision is to be made, we have to sit down and think about the implications this decision will bring about to those yet to come. We are to project our thoughts into future possibilities, weigh the implications, pray, talk, dream and then make a decision.

Another tells us… when we are ready… things will come to be. One story that was related to me by a Haudenosaunee knowledge keeper, Kakwirakeron a.k.a. Arthur Montour of Kahnawake, went like this:

> One way to think about generations is like this. In the bible they tell of a man named Moses who was a Hebrew slave in the kingdom of the Pharaoh in ancient Egypt. One day, God told Moses to lead his people out of bondage and to journey to the promised land.
>
> So the long and the short of it is, Moses did as he was bid to do. He led his people out of Egypt into what is now the Sinai desert. There they roamed. The first generation born in that desert were born free but with the shackles of a slave mentality still held by their relatives… older siblings, mothers, fathers, aunts, uncles, grandparents, great grandparents.
>
> The next generation that followed were too born free, but with the remnants of the slave oppression and indecision from their mothers, fathers, aunts, uncles, grandparents, great grandparents. Generation three is born in the great desert, born of free parents but still shackled by the memories and trauma of

a slave mentality. The fourth generation is born of free parents and grandparents but the memories of trauma are still lingering in the sub conscience of the people. Generations five and six are born into a free society shaking off the last shackles that enslaved their minds. Now this seventh generation that is born are ready to build a free promised land.

This is the process our people must live through in order to become viable nations again.

Personally, I like the chapan perspective.

In the big sky... kicikisik... there swims *Namew*... the sturgeon. Namew swims in the Achak Sipi... river of spirits (Milky Way). Namew is the head fish... *Okimaw Kinisew*. Being that Namew swims in the river of spirits, Namew watches over the ancestors, bloodlines, and generations. Namew is our connection to the future and the past.

We are told we hold sway and influence seven generations. Three generations in front of us... our children... their children and our grandchildren... *noosisimuk* and their children... chapanuk... our great grandchildren. These are the ones that take our hopes, dreams, visions, teachings, fears, biases into the future to those we will never meet. If we want our future generations to be a strong, resilient, loving, inventive, adaptive, kind people then the instruction and example we give our children, grandchildren, and great grandchildren should be truthful, kind, and unwavering. In our past the three generations that got us to where we are now inspired, taught, loved, protected, and guided us, our parents, grandparents, and great grandparents. At each extreme end of this chain of generations we call chapan... "the ones that tie together." One chapan ties together the past and the other ties together the future.

In the sky swimming in the Milky Way directly above the Pleiades are the constellations, Auriga, Perseus, and Andromeda.

Thirteen stars of these three constellations form Namew, the Sturgeon and the seven main stars are the seven generations with us right in the middle, and three generations on either side.

There are those that ask me to talk to them about "Star People"… usually television shows that speculate about "aliens" and "visitors" that came to earth and gave humans the knowledge to build things like the pyramids, great burial mounds, The Nazca Lines in Peru… etc. etc…

I always tell them the human brain does not/did not need any "alien intervention" in order to do what it has done. Creator gave us an awesome tool that has all the answers we seek; *Wiyndip*… the brain.

I go on to say that in regards to the star people… WE are the Star People… not physical visitors from another world or reality. We are the aliens… visiting at this place… at this time… in this form. If there are beings out there that decide to come and visit us then these beings are our Nitootimuk… our relatives. (Heeeeyyy…. CUZZIN!!!) We all come from the same Creator who is still creating… right now as you read this!

I have been invited to appear on television shows such as "Ancient Aliens" Discovery Channel or APTNs' "Red Earth Uncovered" and for some reason or other they just never happened. Guess I was not meant to be there. I know this to be the case because this happens to me despite my most well laid plans about attending something I am oddly curious about or have reservations… yet continue my pursuit of… be it a peyote ceremony, Ancient Aliens interview in Los Angeles, or attempting to see about a Ayahuasca ceremony. I was not meant to be at these places and therefore they did not happen.

Things have been happening like this lately and I am put into a line of thought that says… go slowly… a little at a time… *upasis poko*.

I was all gung ho about planning and holding an international star knowledge symposium where Indigenous star knowledge keepers from all over the world would gather at one spot and share their knowledge about the stars. I envisioned a large field with large ten meter Tipis. Inside these Tipis would be placed inflatable, portable planetariums. I saw ten of such structures.

Inside the planetariums, an Indigenous knowledge keeper from the various Indigenous nations of the world would present on the star knowledge of their people. The sky of the planetariums could be flipped to view whatever sky that was being talked about to a sky familiar to the presenter.

Then COVID-19 happened and everything changed. We switched gears to virtual ZOOM session.

Instead of one large gathering we are now going with three smaller ones to coincide with the equinoxes and solstices with a face-to-face gathering planned for the June Solstice 2021.

Baby steps… slow down… relax… don't panic… have some bannock!

The great singer-songwriter Stevie Nicks wrote about how our days are like web strands flying in the wind and we are forever beginning again.

Kokominakasis reveals to me dreams through her web of possibilities...

We are star people... we begin in the stars and we continue in the stars. My life experience has led me to this place, at this time, and this is what I choose to believe. We are Achak... Spirit... Beings of Light... Beings of Energy... Kisikookuk.

We are in the process of remaking an age old song... singing, dreaming, praying, talking into existence once again, the knowledge of our people...

Ekosi

NAHMEW (the Sturgeon)